OECD Studies on Water

Climate Change, Water and Agriculture

TOWARDS RESILIENT SYSTEMS

This work is published under the responsibility of the Secretary-General of the OECD. The opinions expressed and arguments employed herein do not necessarily reflect the official views of OECD member countries.

This document and any map included herein are without prejudice to the status of or sovereignty over any territory, to the delimitation of international frontiers and boundaries and to the name of any territory, city or area.

Please cite this publication as:
OECD (2014), *Climate Change, Water and Agriculture: Towards Resilient Systems*, OECD Studies on Water, OECD Publishing.
http://dx.doi.org/10.1787/9789264209138-en

ISBN 978-92-64-20912-1 (print)
ISBN 978-92-64-20913-8 (PDF)

Series: OECD Studies on Water
ISSN 2224-5073 (print)
ISSN 2224-5081 (online)

IWA Publishing
ISBN: 9781780406602 (Paperback)
ISBN: 9781780406619 (eBook)

The statistical data for Israel are supplied by and under the responsibility of the relevant Israeli authorities. The use of such data by the OECD is without prejudice to the status of the Golan Heights, East Jerusalem and Israeli settlements in the West Bank under the terms of international law.

Photo credits: Cover: ©Taro Yamada/corbis, © Igor Dudchak - Fotolia.com.

Foreword

Interactions between climate change, water and agriculture are numerous, complex and region-specific. Climate change can affect water resources through several dimensions: changes in the amount and patterns of precipitation; impact on water quality through changes in runoff, river flows, retention and thus loading of nutrients; and through extreme events such as floods and droughts. These changes in the water cycle can in turn deeply affect agricultural production in practically all regions of the world and have destabilising impacts for agricultural markets, food security and non-agricultural water uses. There is thus a strong case for considering agricultural water management and policy in the context of climate change. In the same way, a sound analysis of mitigation and adaptation strategies in the agricultural sector to climate change should place more emphasis on the water cycle.

A sound policy analysis should build on the state of knowledge of the main linkages between climate change, water and agriculture, and to identify the knowledge gaps and uncertainties. An important knowledge gap is related to seasonal impacts, extremes and variability of water availability since many current studies focus on annual timescales. There are challenges in comparing regional impact assessments driven by climate (and other) data from widely differing sources, and this may lead to conflicting and potentially misleading results. Significant uncertainties in future hydrological responses to climate change across models also exist.

The impact of climate change on water quality is also complex and highly uncertain. Increased incidences and severity of flooding could mobilise sediment loads and associated contaminants and exacerbate impact on water systems, while more severe droughts may reduce pollutant dilution, thereby increasing toxicity problems. However, the relationship of water quality to weather and climate is very complex and it is difficult to make projections about it under all climate change scenarios.

From a market and policy perspective, the impact of climate change on the water cycle, and the subsequent consequences for agriculture, highlight the importance of comprehensive adaptation and mitigation strategies. For example, more investment could be required in drainage and water control against floods or water retention against droughts; the expected rise in extreme events may mean that agricultural risk management systems need to be adapted or developed further in some countries, and the scarcity of water resources at certain periods of the growing season could require water sharing arrangements amongst all water users.

Identifying and prioritising adaptation strategies for agricultural water management in the context of climate change are not easy tasks because of the strong uncertainties related to the impact of climate change and the potential of current agricultural systems to cope with these impacts. Moreover, the resilience of agricultural systems to climate change is not solely an issue of water management, although this dimension may dominate in practice. Beyond water efficiency in agriculture, the challenge also resides in building agricultural systems that are less dependent on water resources on the whole. But whatever the relative importance of water in the challenge of adaptation, the fact nevertheless remains that considering agricultural water management without taking into account climate change is not a realistic option.

The main objectives of this report are to review the main linkages between climate change, water and agriculture; to identify and discuss adaptation strategies for a better use and conservation of water resources in the agricultural sector in the context of climate change; and, on the basis of the aforementioned, to provide guidance to help decision makers on an appropriate mix of policies and market approaches to address the interaction between agriculture and water systems under climate change. This report is based on literature reviews on various aspects of the nexus of climate change, water and agriculture.

Acknowledgements

This report, originally started by Kevin Parris (now retired from OECD), has been prepared by Julien Hardelin and Jussi Lankoski of the OECD. It uses material from a set of three consultant reports, in particular:

- *Impact of Climate Change on Water Quantity and Quality and Implications to Agriculture – A Review*, by Ximing Cai (Professor at the University of Illinois at Urbana-Champaign), Xiao Zhang, Paul H.C. Noël and Majid Shafiee-Jood (respectively graduate student, PhD student and Research assistant at University of Illinois at Urbana-Champaign), especially for Chapter 1.

- *Extreme Weather Events and Climate Change: Implications for Agriculture*, by Michael J. Roberts (Associate Professor, Department of Agricultural and Resource Economics, North Carolina State University) and Emiko Naomasa (PhD student, University of Hawaii at Manoa), for Chapters 1 and 2.

- *Agricultural GHG Mitigation Practices, Water Resources and Water Quality*, prepared by Bruce A. McCarl (Distinguished Professor of Agricultural Economics, Texas A&M University), for Chapter 2 on mitigation, and Annex C.

This report was declassified by the Joint Working Party on Agriculture and the Environment in April 2014.

Table of contents

Tables

Figures

Boxes

Executive summary

Water is a central issue of adaptation to climate change in agriculture. Agricultural production depends critically on how climatic variables such as precipitation and temperatures vary across regions and over time. The effects of climate change on agriculture occur through crop water requirements, availability and quality of water, and other factors, which are affected by both long-term gradual change and extreme events, and across a range of scales from local to regional to continental. Moreover, climate is not only changing but is becoming *non-stationary*, meaning that expectations can no longer be based only on past observations.

Interactions between climate change, water and agriculture are numerous, complex and region-specific. Climate change can affect water resources through several dimensions simultaneously: changes in the amount and time patterns of precipitation; impact on water quality through changes in runoff; river flows; retention and thus loading of nutrients; and through extreme events such as floods and droughts. Interactions between relevant weather variables that affect agricultural production, such as temperature and precipitation, are difficult to characterise. Moreover, scientific evidence underlying projected impacts on freshwater has significant limitations when it comes to informing practical, on-site adaptation decisions. These complex interactions multiply the uncertainties concerning the impact of climate change on agriculture.

The frequency and severity of extreme events such as floods and droughts may increase as a result of climate change and have substantial negative impacts on agricultural production. Much of the work undertaken on the potential impact of climate change has focused on projected changes in average temperature and rainfall, and links between these changes and measurable outcomes with clear economic implications. In contrast, there has been generally less confidence about how much night-time temperatures will rise relative to daytime temperatures, how much winter temperatures will rise relative to summer temperatures, whether or how much more variable temperatures will become, and how climate change may change the frequency and severity of extreme rainfall events, tornados and cyclones, etc. Despite the low level of certainty concerning the scientific evidence regarding shifts in extreme events, non-linear (convex) damage functions mean that changes in extremes are expected to be the most costly.

Because agricultural water management involves public goods, externalities and risk management issues, private adaptation to climate change is not equal to collective adaptation. A consistent strategy for agricultural water management needs to consider the following five levels of actions and their linkages.

- *On-farm*: Adaptation of water management practices and cropping and livestock systems.

- *Watershed*: Adaptation of water supply and demand policies in agriculture and with the other water users (urban and industrial) and uses (ecosystems).

- *Risk management*: Adaptation of risk management systems against droughts and floods.

- *Agricultural policies and markets*: Adaptation of existing agricultural policies and markets to the changing climate.

- *Interactions* between mitigation and adaptation of agricultural water management.

There is room for public policies that create an enabling environment for on-farm adaptation. Depending on the region, adaptation of agriculture may require only marginal adjustments, such as earlier sowing dates, or deep structural changes involving complete changes of production patterns, and update of cropping and livestock systems. These structural changes would be at the origin of substantial adjustment costs and may not be affordable by all farms, in particular those which already face severe financial constraints. Public policy interventions can facilitate on-farm adaptation by collecting and disseminating relevant and site-specific information about projected impacts of climate change and best adaptation practices and providing technical assistance. In cases of structural change, public policy could help smooth switching costs across time by adaptation planning and offering temporary financial assistance in clearly defined circumstances.

At the watershed level, well-designed, flexible and robust water sharing rules and economic instruments such as water pricing and water trading can foster adaptation of water systems. As climate is becoming non-stationary and climate risks are projected to increase, systems that allocate water across farms and across other uses should be flexible and robust enough to allow for efficient use of water, taking into account redistributive consequences and priority uses. The shadow price of water can vary a great deal within the growing season. Adaptation is a long-term, continuous process that involves learning, investment, and may be affected by path dependency. Two types of incentives for improving water allocation should be considered depending on the time horizon.

- *Short-run incentives* that allow farm systems to cope with intra-seasonal volatility of water supply and reallocate water to its most efficient uses within the growing season.

- *Long-run incentives* to adapt to continuously changing conditions of water supply, taking into account other factors (growing population, increase in urban demand, ecosystems, etc.).

The relative strength of economic instruments should be considered in light of these two time horizons. Past and current country experiences with these economic instruments constitute a strong basis for further development in the context of climate change. Water pricing and water trading systems could not only provide a stable economic incentive for investment in adaptation, but also valuable flexibility in the short run to adjust to seasonal variations of water supply and demand. Flexibility in the design of the system would send the proper signal of scarcity to water users and thereby allow the quantity or price of water to be adjusted accordingly. However, a major challenge is that it may be politically costly to reduce the size and number of quotas, or to increase the price of water in order to reflect projected decreases of average water availability. More generally, the issue of long-run incentives requires adopting adaptation planning based on the best scientific evidence available and to undertake careful consideration of the already existing set of water rights, whether these are explicit or implicit.

Risk management instruments such as prevention and insurance can play a major role for managing the increased risk of floods and droughts and contribute to the resilience of agriculture to climate change. Climate change is deeply reshaping the landscape of natural risk management. Even without non-stationary climate change, weather extremes such as floods and droughts were already not easy to insure for several reasons, including correlated risks and lack of statistical information for risk pricing. Innovative risk management tools, such as weather index insurance and catastrophic bonds, can potentially play a role by improving the efficiency of insurance systems but are still at an experimental stage. Moreover, it remains important to have a holistic approach of risk management in agriculture.

Commodity markets can play an important role in smoothing the impact of extreme weather events on price volatility over time. Open trade is an important vehicle to fully reflect shifting comparative advantage due to climate change while also pooling the risk so that yield losses in a given region can be offset through imports. Well-functioning competitive storage markets can reduce the cost of inter-temporal price volatility related to extreme weather events and reduce the probability of food price spikes on a temporary basis. The way storage markets can integrate non-stationary climate requires, however, further research. Government intervention could have a potential role in providing the enabling environment to promote private storage and competitive storage markets.

Climate change mitigation practices may have positive or negative implications on agricultural water management and on water quality. The potential synergies and trade-offs between mitigation and agricultural management practices are, however, site-specific and for many cases there are substantial knowledge gaps. Although this is a complex matter, it is important to recognise these linkages in the design of mitigation policies to reduce the risk of conflict between mitigation and water policy objectives and to maximise potential synergies.

Chapter 1

Impact of climate change on the water cycle and implications for agriculture

The present chapter provides a review of the main impacts of climate change on water supply and demand for agriculture, including the crop and livestock sectors, on the relations between climate change, agriculture and water quality, and on extreme water events such as droughts and floods. The principal aim of this chapter is to provide a background for the policy analysis of climate change adaptation and mitigation strategies.

Climate change is likely to affect farm systems through several channels, which interact in complex ways involving direct and indirect impacts. An increase in the atmospheric CO_2 concentration, all other things being equal, can in theory boost crop growth and yields through increases in the photosynthesis process (CO_2 fertilisation). But it can also affect climatic variables such as temperatures, rainfall, and vapor pressure deficit in a complex way. The impact of climate change on a given farm system will be ultimately the outcome of a *combination of interacting climatic variables* at the *local scale*. Moreover, *seasonal patterns* and *variability* of climatic parameters could be as important as changes in average projections of climate variables. The impact of climate change on agriculture thus combines a *cascade of multiplicative uncertainties* (OECD, 2010; OECD, 2013a).

- Uncertainty about climate change projections, including the magnitude of climate change, temporal pattern of climate variables, and their interactions.

- Uncertainty about the biological responses of cropping and livestock systems, etc.

- Uncertainty about the potential benefits and costs of adaptation responses to climate change.

Even if interactions are complex and uncertainties high, as a major input for agricultural production, water necessarily plays a major role in terms of impact of climate change on agriculture and in the adaptation process. **Figure 1.1** proposes a simplified framework that summarises the channels through which climate change may affect farm systems through changes in the water cycle (a more detailed analysis of the linkages can be found in Annex A from FAO, 2011). Climate change is projected to lead to long-term changes on climate variables such as temperature, precipitation, etc. At the same time, climate change could also lead to an increase in the frequency and severity of extreme climate events such as floods, droughts, and cyclones. Climate change can affect both the supply and demand for water in agriculture.

- Water supply can be *directly* impacted by climate change through changes in rainfall patterns, and *indirectly* through changes in water compartments such as surface water, groundwater, snow and glaciers that can be used for the purpose of agricultural water withdrawals, including irrigation and livestock.

- Water demand by agriculture can be modified due to changes in cropping and livestock system for adaptation purposes and changes in crop water requirements driven by climatic variables such as high temperature and winds.

To these multiple sources of uncertainty should be added the *role of non-climatic drivers* that can also affect water supply and demand, including population growth, changes in diet and lifestyle in emerging and developing countries, and growing urbanisation. According to the *OECD Environmental Outlook to 2050* (OECD, 2012a), the baseline scenario projects a 55% increase in global water demand, from 3 500 km^3 in 2000 to 5 500 km^3 in 2050, most of this increase comes from manufacturing, electricity and domestic use (**Figure 1.2**).

Characterising the impact of climate change on agriculture, including the specific role of water, requires taking into account the adaptation capacities of the sector. There is a typical distinction between *potential impacts* of climate change, that do not include adaptation responses or capacities in the analysis, and *net impacts*, that include these dimensions. For instance, suppose that climate change increases the variability of water supply (e.g. rainfall, groundwater) in a given region. There is then a potential impact in terms of increasing yield risk. But the net impact will depend on the ways farmers will be able to deal with this water supply by, for example, improving the productivity of cropping systems, using drought tolerant varieties or increasing the efficiency of water management at the farm-level. To be able to measure the net impacts of climate change requires taking these adaptation responses into account. **Figure 1.3** presents the three major components that need to be considered in any assessment: exposure and sensitivity that define potential impacts, and adaptation responses, which are "subtracted" from potential impacts to define the net impacts.

Figure 1.1. Main linkages between climate change, the water cycle, and farm systems

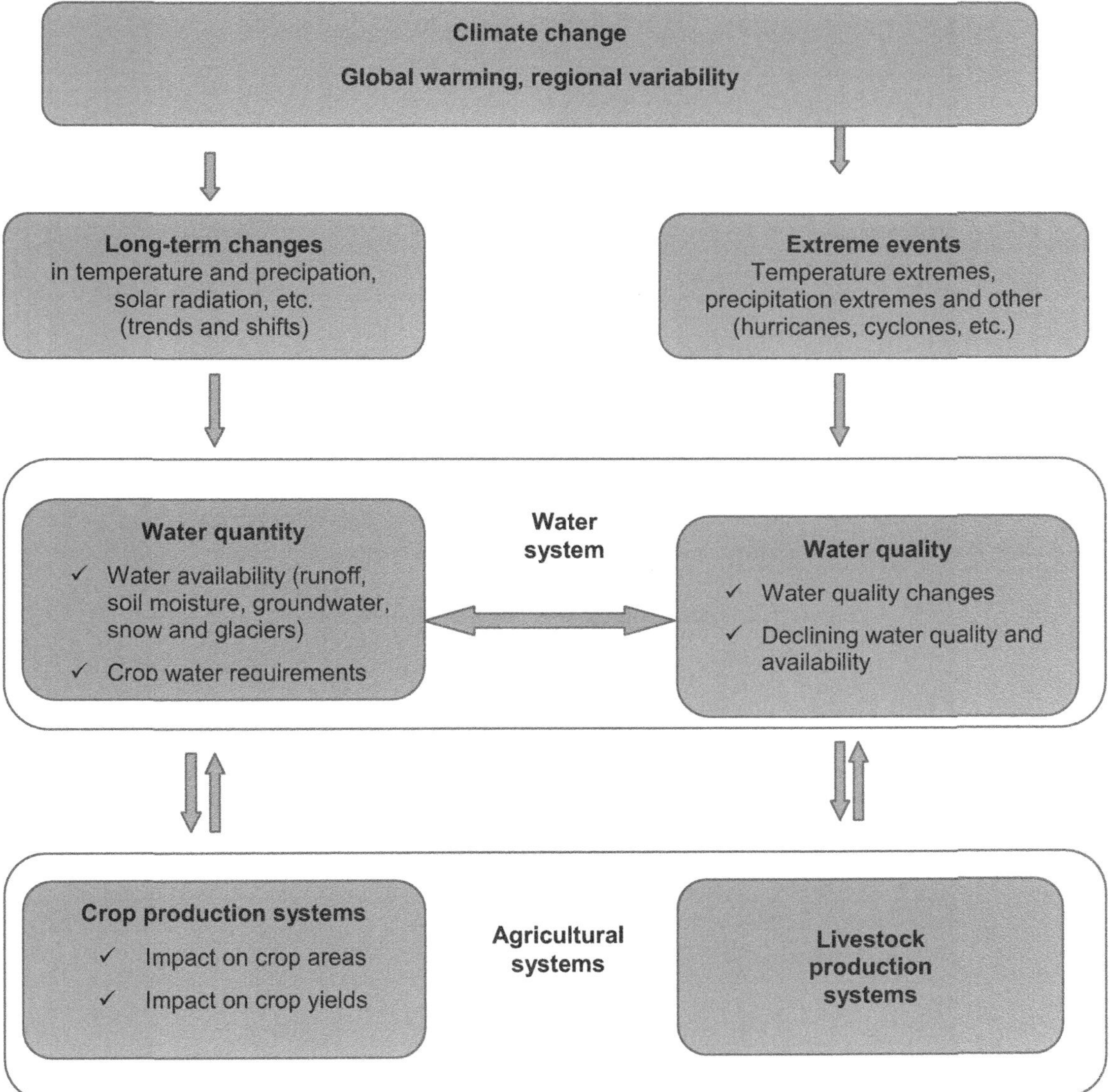

Source: Cai, X., X. Zhang, P. Noël, and M. Shafiee-Jood (2013), "Impact of climate change on water quantity and quality and implications to agriculture: A review", unpublished consultant report.

Figure 1.2. Global water demand: Baseline scenario, 2000 and 2050

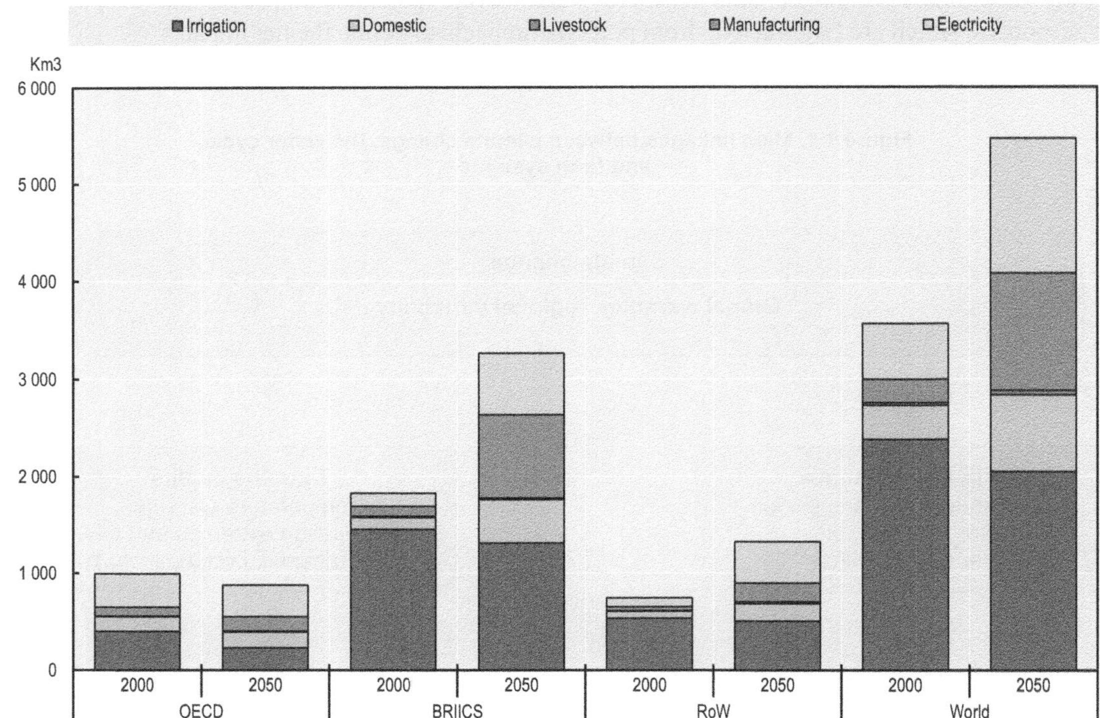

Note: BRIICS: Brazil, Russia, India, Indonesia, China, South Africa; RoW: Rest of the World.

Source: OECD (2012), *OECD Environmental Outlook to 2050: The Consequences of Inaction*, OECD Publishing. doi: 10.1787/9789264122246-en.

Figure 1.3. Potential and net impacts of climate change

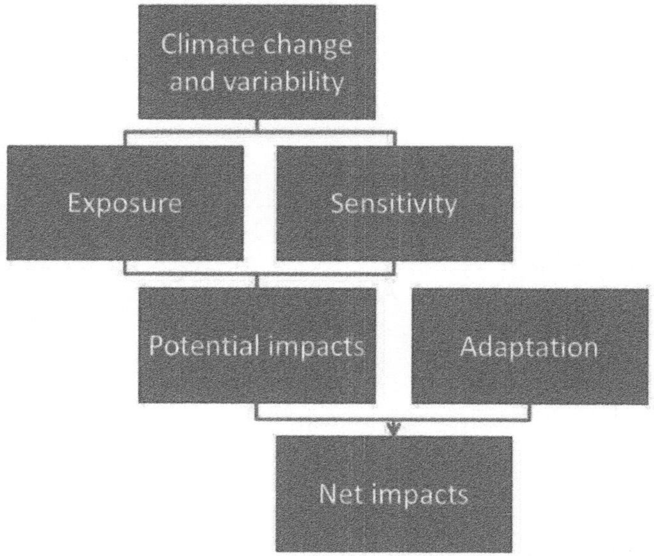

Source: Adapted from Fellmann, T. (2012), "The assessment of climate change-related vulnerability in the agricultural sector: reviewing conceptual frameworks", in A. Meybeck, J. Lankoski, S. Redfern, N. Azzu, and V. Gitz (eds.) *Building resilience for adaptation to climate change in the agriculture sector – Proceedings of a Joint FAO/OECD Workshop*, Rome, 23-24 April 2012.

Impact of climate change on the water cycle

Precipitation and surface water

Projections of precipitation patterns remain highly uncertain, especially when looking at local scales (OECD, 2013a). On the whole, projections from climate models indicate that climate change could lead to an acceleration of the water cycle, involving more frequent and intense rainfall episodes. Beyond this general tendency, climate change is likely to have impacts of a different nature on the water cycle in different regions of the world. Notably, available climate projections indicate likely increases in mean precipitations at high latitudes as well as at mid-latitude wet areas, and decreases in summer precipitation in mid-latitude and sub-tropical dry areas (IPCC, 2013). In regions with rain-dominated catchments, enhanced flow seasonality was concluded by various studies, implying higher peak flows and decreased low flows and extended dry periods (IPCC, 2007a; EEA, 2012).

Such modifications of precipitation patterns would affect the hydrological dynamics of water compartments such as rivers, lakes and groundwater, and glaciers. Increases in precipitation in winter and spring could translate into higher river flows, with eventually more frequent and severe flood events. Alternatively, lower precipitation in summer is likely to reduce river flows, with potential risks of temporary water shortages during critical phases of crop growth. Such shifts in the seasonality of river flows are, for example, projected in most parts of Europe, except for most northern and southern regions. **Figure 1.4** presents an illustrative example of projected changes in river flows in the case of four European rivers: the Rhone (Switzerland and France), the Danube (Central Europe), the Indalsaelven (Sweden) and the Guadiana (Spain).

Figure 1.4. Projected change in daily average river flow:
Rhone, Danube, Indalsaelven and Guadiana

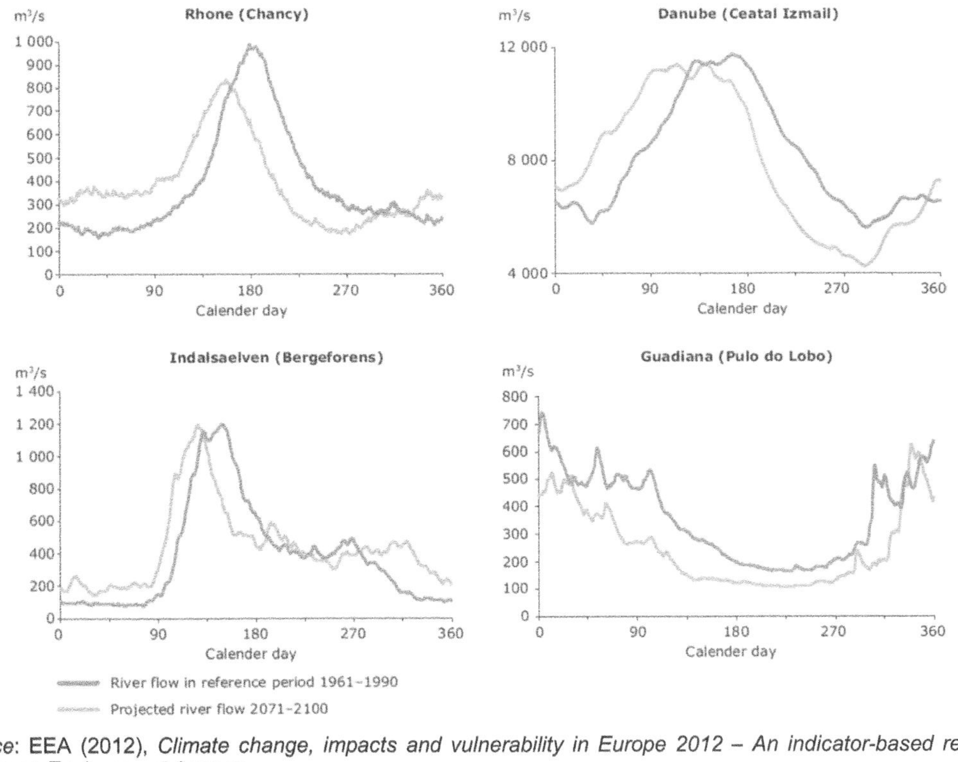

Source: EEA (2012), *Climate change, impacts and vulnerability in Europe 2012 – An indicator-based report*, European Environment Agency.

Glaciers and snow

Glacial ice is the largest reservoir of freshwater on earth. Water stored as glacial ice is a region's hydrologic insurance (e.g. during the 2003 European drought, the flow in the Danube River was three times higher than the long-term average flow). It is believed that glacial mass has been affected by long-term climate changes, with a trend towards glacier retreat, especially since the 1980s. Glaciers play also a major role in several parts of the world to smooth water flows in rivers across time, for which irrigation is in some cases strongly dependent on (FAO, 2011).

Groundwater sources

Changes of precipitation patterns, both intra-annual and inter-annual, are also likely to affect groundwater recharge rates, although there is a lot of uncertainty, as well as local variations related to these impacts (Taylor et al., 2012). Climate change could also have implications in terms of the interactions between ground and surface water compartments, as water stress on the latter could be transmitted to the former in the dry season. One can also expect an indirect effect of climate change through an increasing demand for groundwater as surface water becomes scarce and droughts more frequent and severe. Groundwater depletion may occur in some places and be aggravated in places where depletion already exists. Agricultural water withdrawal, including irrigation, already represents a substantial amount of groundwater withdrawals in some countries (**Figure 1.5**), and a trend towards more frequent droughts may be an incentive to close the crop water requirement gap with groundwater resources.

**Figure 1.5. Share of agricultural groundwater use in total groundwater use,
and total groundwater use in total water use, OECD countries, 2007**

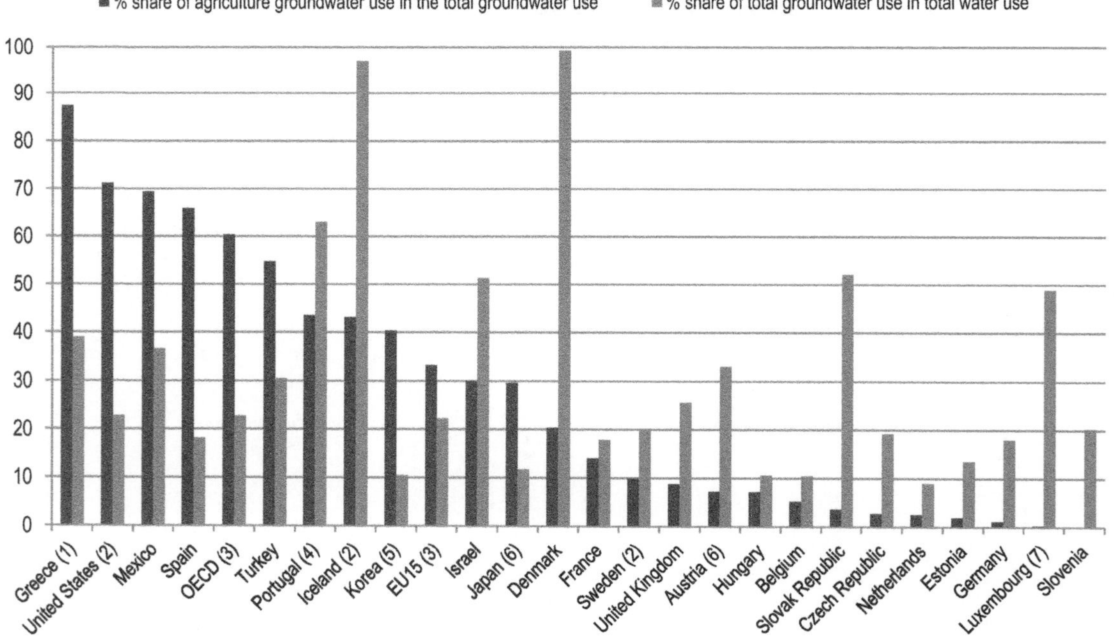

1. Data for Greece refer to year 2004.
2. Data for United States and Iceland refer to year 2005. In Iceland, agriculture groundwater use includes fish farming.
3. The EU15 and OECD data must be interpreted with caution, as they consist of totals using different years across countries, and do not include all member countries. EU15 excludes: Finland, Ireland and Italy. OECD excludes: Australia, Canada, Chile, Finland, Ireland, Italy, New Zealand, Norway, Poland and Switzerland.   4. Data for Portugal refer to year 2000.   5. Data for Korea refer to year 2002.
6. Data for Japan and Austria refer to year 2008.   7. Data for Luxembourg refer to year 2010.

Source: OECD (2013b), *OECD Compendium of Agri-environmental Indicators*, OECD Publishing.doi: 10.1787/9789264186217-en; OECD Agri-environmental Indicators Questionnaires.

Water availability and agricultural production under climate change

Impacts on crop water requirements

Climate change can affect not only water availability, but also crop water requirements. Crop water requirements are defined as the "quantity of water required by a crop in a given period of time for normal growth under field conditions" (FAO, 2008). In cases where precipitation is lower than the evapotranspiration requirement, there is a *water deficit*. In such situations, irrigation water can fill the gap. Under climate change, current rainfed crops may require irrigation to maintain reasonable productivity, and current irrigated crops may have a larger or smaller irrigation requirement (FAO, 2011).

A few studies have investigated the climate change impacts on global agricultural water requirement (Döll, 2002; Fischer et al., 2007). Döll (2002) predicted a slight increase in global crop water requirements, but significantly less than other comparable works which tend to estimate an increase by 5-20% (Fischer et al., 2007; Nelson et al., 2009). Recently, climate projections to 2070-2099 by Zhang and Cai (2013) indicate contrasting results across countries: lower water requirements can be expected in some regions of the world such as **Africa**, **Australia** and **China**; while in regions such as **Europe**, **North India**, eastern **South America** and eastern **United States** higher crop water requirements are projected. These results are subject to the land use types (rainfed or irrigated) and the uncertainty involved in the assessment approaches.

More specific results are provided for the different regions, although the results are subject to uncertainties in climate change predictions (Holden et al., 2003; Tao et al., 2003). Tao et al. (2003) found that agricultural water requirement might increase in north **China** while decrease slightly in south **China** in 2020s. The requirements and deficits could decrease in Western **Europe** but increase in Eastern **Europe** (Zhang and Cai, 2013). Southern **Europe** has a higher probability of experiencing greater water gaps than the north due to the altered precipitation (Bates et al., 2008). Water stress is likely to intensify in the Mediterranean (**Portugal**, **Spain**) and some regions of **Central and Eastern Europe** (Döll, 2002). Substantial irrigation requirements may be expected in some countries where irrigation hardly exists at present (e.g. Ireland, Holden et al., 2003).

With most its cultivated land located along the coastline, **Australia** is more vulnerable to issues caused by a rise in sea level, more frequent and intense storms, monsoon, and other changes. When considering only the joint effect of temperature and precipitation changes and assuming constant land use type, the irrigation requirements and water deficits on rainfed areas may decrease by the end of this century, although the magnitudes are subject to high uncertainty (Zhang and Cai, 2013). Quiggin and Horowitz (2003) found that temperate agriculture in Australia might gradually move southwards where the climate would become more suitable for agriculture. However, Kingwell (2006) argued that the spatial complexity might affect the "moving south" story.

North America is spatially heterogeneous and the effects of climate change also vary significantly with space and time. The recent study by Zhang and Cai (2013) finds contrasting results depending on the region considered. Results on water stress by region can also vary a lot between climate scenarios and models. Seasonal drying is likely to occur in some regions, even if more precipitations can also be expected (FAO, 2011). If the drying occurs in the critical period for plant growth, then irrigation requirements would be largely affected.

The change of the timing of the growing season for specific crops also complicates the estimate of irrigation requirements under climate change (Minguez et al., 2007; Ortiz-Bobea, 2012; Ortiz-Bobea and Just, 2012). Rising temperature would extend the length of the growth period, allowing both earlier planting and later harvesting in the northern temperate zones, but may at the same times shorten it in other regions of the world (Piao et al., 2006; FAO, 2011).

Longer growth periods will likely increase crop water requirements. Furthermore, one crop may become unsuitable to cultivate in a particular region as the climate changes. Such agricultural alteration makes the estimates of crop water requirement more complicated.

Impact on crop yields

Climate change affects crop yields through a variety of climatic channels: changes in atmospheric CO_2, temperature increase, altered precipitation and transpiration regimes, increased frequency of extreme water events, pests, and weeds (Tubiello et al., 2007). These climatic variables interact with agricultural management practices, and more broadly cropping and livestock systems. It is difficult to insulate the proper impact of a change in water variables on crop yields, but it is likely that water could become a major limiting factor for crop production in most regions of the world. There are two principal methods to assess the impacts of climate change on crop yields: agronomic models and statistical analysis (**Box 1.1**), which differ in the way they include the influence of water input on crop yields.

Using the statistical yield approach, Lobell et al. (2011a) found that global production of maize and wheat is expected to reduce by 3.8% and 5.5%, respectively, due to climate change; while for soybeans and rice, positive and adverse effects offset overall. The primary reason for such effects is that the temperature increasing trend from 1980 to 2008 was large enough to offset the benefits brought by technology, CO_2 fertilisation and other factors. In another study, Lobell and Field (2007) showed that "despite the complexity of global food supply, (...) simple measures of growing season temperatures and precipitation – spatial averages based on the locations of each crop – explain ~30% or more of year-to-year variations in global average yields for the world's six most widely grown crops. For wheat, maize and barley, there is a clearly negative response of global yields to increased temperatures."

Box 1.1. Approaches to assess the impacts of climate change on crop yields

Agronomic models

The earliest and most prevalent approach for assessing climate change impacts on agricultural outcomes, including the last 2007 IPCC report, makes use of *deterministic crop models* developed by plant physiologists. These mathematical models rely on fundamentals of photosynthesis combined with the developmental processes of plant growth and seed formation. Inputs into the models are generally daily measures of temperature, rainfall and solar radiation. Crop growth is ultimately limited by the amount of solar radiation, available nitrogen, or available water. Examples of such models include CERES (Crop Environment Resource Synthesis) and STICS (Simulateur multidisciplinaire pour les Cultures Standard). A large number of these models are now compiled and maintained in a software program called DSSAT (Decision Support System for Agrotechnology Transfer). Furthermore, the Agricultural Model Intercomparison and Improvement Project (AgMIP) aims at improving crop and economic projections by linking and comparing existing modelling approaches (Rosenzweig et al., 2013).

Soil characteristics enter the model mainly with regard to how much moisture they can retain and store from rainfall events. Soil characteristics can also be important for nutrient content (particularly nitrogen for grains), if these nutrients are not applied in appropriate quantities by farmers. In developed countries, lack of nutrients rarely limits crop production. In developing countries, however, nutrient limitations are a major consideration.

Overarching challenges with crop models are that they are often calibrated using data from experimental plots, and their record for predicting yields of actual farmers is not well understood. Over time, some models have become more complex, progressively incorporating adjustments and additional factors in an effort to improve predictive accuracy and hypotheses and laboratory evidence that might explain differences between modeled and observed outcomes.

continued

Statistical models of crop yields

Another approach to modeling crop outcomes in relation to weather and climate is to draw on direct statistical comparisons between weather, climate and observed yield outcomes (Roberts et al., 2012). An important feature of the statistical approach is the heterogeneity of weather and climate. Looking over time in a fixed location, weather variations are effectively random from the point of view of a farmer. Comparing crop outcomes to climate across space is a more delicate matter as local differences besides climate may well be associated with climate, leading to non-causal associations.

Where time-series evidence serves as a viable natural experiment, relying on random weather variation, cross-sectional comparisons – if adequately defended against potential omitted variables bias – may provide evidence on the scope for adaptation.

There are, however, challenges to the statistical approach of crop yields, notably: correlations, even if they can be defended as causal in a statistical sense of weather or climate affecting yield outcomes, do not reveal the physiological mechanism that underlies the association; looking at a specific crop has limitations because one form of adaptation would be to choose a different crop; and CO_2 fertilisation is by nature not taken into account.

The overarching conclusions from statistical models of crop outcomes are that extreme heat tends to be a more powerful predictor of yield than precipitation (Lobell and Burke, 2008). This empirical pattern may appear contrary to the essential predictions of crop models, which often point to precipitation as the main limiting factor for non-irrigated agricultural systems (Sinclair, 2010). Recently, there has been a growing effort to reconcile this apparent disconnect between crop models and statistical models.

Figure 1.6. Projected change in water-limited crop yield in Europe in 2050

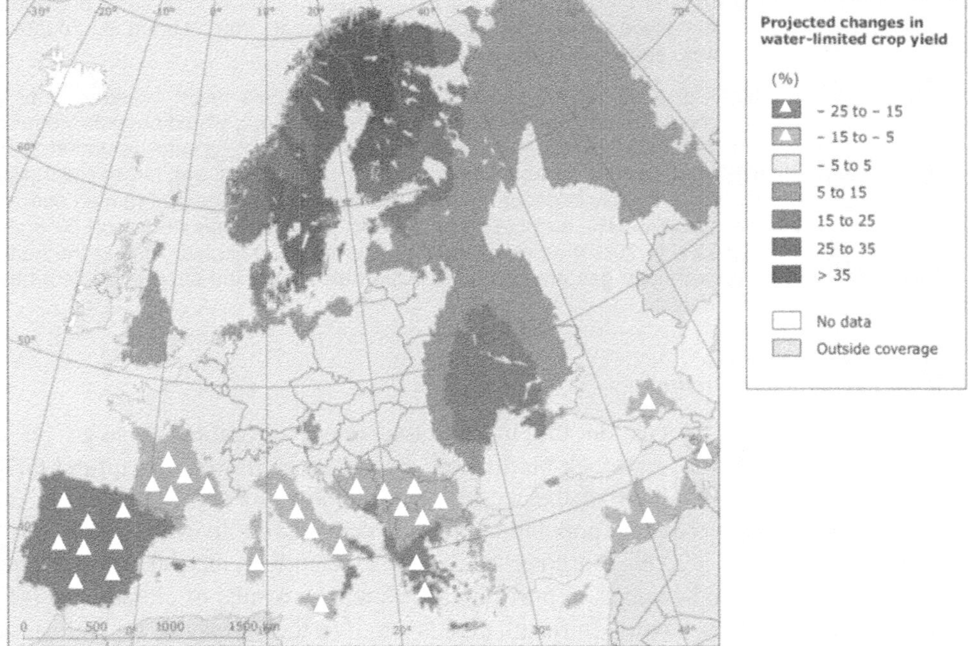

Mean relative changes in water-limited crop yield simulated by the ClimateCrop model for the 2050s compared with 1961–1990 for 12 different climate models projections under the A1B emission scenario.
Source: adapted from Iglesias, A., Quiroga, S. and A. Diz (2011), "Looking into the future of agriculture in a changing climate", *European Review of Agricultural Economics*, Vol. 38(3).

The recent assessment of the impact of climate change and vulnerability in **Europe** (EEA, 2012) also provides some estimates of the projected changes in water-limited crop yields for Europe in 2050 (**Figure 1.6**). Crop yields could increase by more than 25% in the northern European countries such as **Sweden** and **Norway**, while significant reductions are projected in the southern area: between -15% to -25% in **Spain** and **Greece**, between -5% and -15% in the south-west of **France** and **Italy**. This illustrates in the case of Europe the extent to which changes in the water cycle arising from climate change are susceptible of redrawing the map of agricultural production in the world. Other examples of regional studies are presented in **Box 1.2**.

Box 1.2. Examples of regional studies on crop yields

In **China**, warming is likely to be detrimental to rainfed crops but beneficial to irrigated agriculture (Wang et al., 2009a). For instance, rice yields in the northeast have increased by 4.5–14.6% per °C due to nighttime temperature increase during 1951–2002 (Tao et al., 2008). By contrast, wheat yield (6–20% per °C) decreased as a result of warmer daytime temperatures (Tao et al., 2008). Increased temperature is generally beneficial for crop yield in the temperate climate zones of northern China (Piao et al., 2010). A study on China found that rice, maize and wheat yields could be reduced by 18–37% in the next 20–80 years under climate change without CO_2 fertilisation effect (Lin et al., 2005). However, the negative impacts of droughts and floods are climbing and have caused significant losses (Piao et al., 2010).

India is likely to be negatively affected by rising temperature and constant or less precipitation (IFPRI, 2009). One study for India found that the crop yield projection over the medium term might decline slightly (4.5-9%), while the long-term projection sees that a substantial reduction (25% or more) is likely to occur (Guiteras, 2007).

In **Australia**, several studies have shown that increasing carbon dioxide levels enhance wheat growth as a result of increased photosynthesis and water use efficiency (Reyenga et al., 1999; van Ittersum et al., 2003; Howden and Jones, 2004; Luo et al., 2005a,b; Ludwig and Asseng, 2006; Anwar et al., 2007; Crimp et al., 2008; Wang et al., 2009b). However, these authors also showed that predicted growth increases from carbon dioxide will be more than offset by decreases in yields as a result of higher temperatures and lower rainfall. Higher temperatures have also been predicted to downgrade grain protein levels (Reyenga et al., 1999; van Ittersum et al., 2003; Luo et al., 2005b) and reduce the length of the growing season (Sadras and Monzon, 2006).

In the **United States**, Cai et al. (2009) showed that Central Illinois might expect a drier and warmer summer during the corn growing season and other times of the year would become wetter and warmer in the Midwest of the United States. According to these authors, "greater temperature and precipitation variability may lead to more variable soil moisture and crop yield, and larger soil moisture deficit and crop yield reduction are likely to occur more frequently (…).The expected rainfed corn yield in 2055 is likely to decline by 23%–34%, and the probability that the yield may not reach 50% of the potential yield ranges from 32% to 70% if no adaptation measures are instituted. Among the multiple uncertainty sources, the greenhouse gas emissions projection may have the strongest effect on the risk estimate of crop yield reduction."

Impacts on livestock production

The previous sections focus on the impact assessment of climate change on crop production systems. However, livestock production accounts for 40% of agricultural GDP in the world and employs 1.3 billion of the world's population (Seo and Mendelsohn, 2008; FAO, 2011). With the increasing population, global demand on livestock products will increase (Thornton et al., 2009; Nardone et al., 2010; Henry et al., 2012). Anticipated increases in temperature and changes in precipitation pattern as results of climate change and variability are expected to have significant effects on livestock systems as well as crop systems. In particular, changes in frequency and severity of extreme events such as heat stress, droughts and floods (as discussed in the extreme event section) will affect livestock productivity, especially in regions particularly sensitive to such events (e.g. Africa). Nevertheless, the literature is limited on this topic. There is a general lack of knowledge about the interactions between climate and livestock, except for certain limited and local areas (Kabubo-Mariara, 2009; Thornton et al., 2009; Nardone et al., 2010). This section discusses some of the results of existing studies and literature, showing how far scientists have gone to assess the impacts of climate change on livestock. The review extends beyond the scope of climate change impact through water; however, all of the impacts on livestock are directly or indirectly connected to climate change impacts on water resources.

Climate change effects on livestock

Industrialised livestock systems contribute the most to animal production in developed countries. These systems are more affected by indirect impacts of climate change (e.g. soil infertility, water scarcity, grain yield and quality, and diffusion of pathogens) rather than direct impacts (Nardone et al., 2010). In these systems, advanced technologies and management actions can enable livestock to better cope with unpleasant conditions. Grazing and mixed-farming systems could be more directly impacted by climate change due to their higher dependence on weather conditions; but on the other hand they may be less vulnerable and more resilient to shocks due to the diversification of farm activities, which can play a role of self-insurance. In arid or semi-arid rangelands, productivity is likely to decline, drought is likely to become more frequent and the natural resources degradation is likely to speed up as in Africa and Central Asia. In the Near East, where rangeland is the dominant land type, productivity is also likely to decline due to the expected decrease in available moisture (FAO, 2011).

Climate variability and extreme climatic conditions affect livestock growth, animal production, and economic efficiency of animal husbandry (AIACC, 2006a; AIAAC, 2006b). Most of the studies in the literature have tried to address the possible ways through which climate change may affect livestock performance and production. It was found that climate conditions (e.g. temperature, humidity, wind speed and precipitation) can affect animal performances including weight growth, milk production, wool production and reproduction directly or indirectly through quality and quantity of feedstuffs and also severity and distribution of livestock diseases (Seo et al., 2010). Based on the literature (e.g. Rötter and van de Geijn, 1999; Kabubo-Mariara, 2009; Thornton et al., 2009; Singh et al., 2012), the impacts of climate change on livestock can be classified into the following categories:

- ***Quantity and quality of feeds***: Pasture yields are strongly affected by weather and climatic conditions, especially water availability. Pasture is the main source of feed intake for livestock in many regions. For example, in Mongolia, livestock obtains over 90% of its annual feed from annual pastures (AIACC, 2006b). Changes to pasture quantity and quality (reduced nutrition) due to changes in temperature and water conditions may affect animal reproduction rate (Harle et al., 2007).

- ***Heat stress***: Heat stress is one of the most important factors affecting animal production (Rötter and van de Geijn, 1999; Frank et al., 2001). Heat stress caused by rising temperature and increased humidity may lead to: significant changes in feed intake, decline in productivity, reduction in milk yield and meat quality, decrease in reproduction efficiency, increase in energy deficit which decreases fertility, fitness and longevity, and in extreme cases mortality as reported in the United States and northern Europe (Rötter and van de Geijn, 1999; Parsons et al., 2001; Harle et al., 2007; Thornton et al., 2009; FAO, 2011; Henry et al., 2012).

- ***Water demand***: Water restriction will worsen negative aspects of high heat load (Henry et al., 2012). Thus, livestock will tend to stay closer to watering points and increase grazing pressure in these regions possibly contributing to land degradation (Harle et al., 2007). There are few studies in the literature about how livestock water demand would change in response to climate change. However, because livestock satisfy a portion of their water need from the water stored in forage, it is difficult to quantify the amount of their water needs from different sources, as forage water content varies with climate also (Thornton et al., 2009).

- ***Livestock health***: Increased temperature together with rainfall pattern changes increases the incidence of pests and diseases (Harle et al., 2007; FAO, 2011; Henry et al., 2012). However, it is very difficult to predict when and where diseases are likely to occur under climate change. Thus, transmission mechanisms are often oversimplified (Thornton et al., 2009). Also it should be noted that although warming may adversely affect livestock production during warm periods, it will probably be beneficial during cold periods. It is still uncertain whether potential future temperature increases will be within the range that livestock can tolerate. Likewise, it is still uncertain whether livestock can cope with an increased frequency of extreme heat stress (Thornton et al., 2009). The literature has not yet studied the impact of climate change on livestock health in depth (Nardone et al., 2010).

Table 1.1 presents some results from a selection of regional studies on the impact of climate change on livestock.

Table 1.1. Selected regional studies on the impact of climate change on livestock

Region of the world	Projected impact of climate change on livestock
Africa	• Livestock sector vulnerable to climate change due to technological gaps and lacking infrastructure (Kabubo-Mariara, 2009). • Reduction of beef cattle production because of drier climate (Seo and Mendelsohn, 2008).
Asia	• Mongolia: Increasing frequency of extreme events, such as "dzud", may affect livestock mortality rates (AIACC, 2006b). • India: More frequent drought and heat waves due to climate change may increase losses in the cattle and buffalo sector (Singh et al., 2012).
North America	• Potential negative impacts of climate change on livestock production through changes in forage productivity and quality; milk production may decrease (Rötter and van de Geijn, 1999: Frank et al., 2001). • Drought and heat waves, expected to increase in frequency and severity, may cause substantial production losses, as past experiences can suggest.
Australia	• Livestock sensitive to climate change although with geographical variability (Howden et al., 2008), in particular through the impact of climate change on feed production related to extensive grazing (Henry et al., 2012). • Adaptation (genetic improvements, farm practices, etc.) can substantially reduce the impacts of climate change on grazing sytems (Seo and McCarl, 2011; Henry et al., 2012).
South America	• High exposure due to the economic importance of the livestock sector in several South American countries such as Argentina and Brazil (Seo et al., 2010).

Source: OECD Secretariat, based on Cai, X., X. Zhang, P. Noël and M. Shafiee-Jood (2013), "Impact of Climate Change on Water Quantity and Quality and Implications to Agriculture – A Review", unpublished consultant report.

Water quality and agriculture under climate change

Water quality is a part of both water availability for human society and the environment; water quality and quantity are connected and affect each other. However, the impact of climate change on water quality has not been sufficiently studied compared to water quantity (Kundzewicz et al., 2007; Bates et al., 2008; OECD, 2012b), although the issue was brought up many years ago (e.g. Murdoch et al., 2000). Two reviews by Whitehead et al. (2009) and Delpla et al. (2009) provide the present state-of-the-art of assessing and predicting water quality effects under climate change. The direct impacts of climate change on water quality include the following.

- Warming of water compartments such as rivers, lakes, etc. This may in turn affect chemical and biological process; in particular, some climate change induced conditions, such as drought, may favour water acidification (Wilby, 1994; Dillon et al., 1997; Whitehead et al., 2009).

- Runoff could be affected by more frequent and severe extreme weather events such as intense precipitation and floods. This may cause soil erosion and affect the mobility and dilution of contaminants, the morphology of rivers and the transfer of sediments in rivers.

As pointed out by Delpla et al. (2009), beyond the direct impact on water quality due to changes of climatic variables, water pollution is linked to urban, industrial or agricultural activities, and climate change could affect water quality indirectly through these activities. Water users such as agriculture, urban areas and industry will have their own adaptation and mitigation strategies in response to climate change, which could in turn affect water quality. For example, extension of irrigated areas in some regions, which might be an adaptation option for the farm sector, could in turn increase irrigation freshwater withdrawals and reduce water flows of rivers below minimum environmental flows. On the contrary, improvements in irrigation water efficiency can have co-benefits as regards minimum environmental flows. Another example is pesticide use: in regions where climate change increases pest risks, this may lead to an increase in pesticide use per unit area, and thus affect water quality. These examples show that the issue is twofold: is agricultural water availability likely to be constrained by reduced water quality under climate change? Will water quality decline due to changed agricultural activities?

Agricultural water availability constrained by reduced water quality

Water use in agriculture can be constrained in two major ways through changes in water quality arising from climate change: water salinisation and soil erosion due to intense rainfall.

Water salinisation in coastal regions can affect agricultural production (Sherif and Singh, 1999; Kundzewicz et al., 2007; Sonnenborg et al., 2012; Werner et al., 2012). According to Kundewizc et al. (2007), "saline intrusion due to excessive water withdrawals from aquifers is expected to be exacerbated by the effect of sea-level rise, leading to reduction of freshwater availability." This can be an important issue as one-quarter of the global population lives in coastal regions where less than 10% of global renewable water resources is available for human uses (Kundzewicz et al., 2007).

More intense rainfall and the rise in extreme weather events are susceptible to cause soil erosion, and thus reduce soil fertility for agricultural production, and have negative impacts on water quality (Bates et al., 2008).

Water quality changes due to changed agricultural production practices

Climate change will eventually entail changes in agricultural land use and in agricultural management practices that in turn can have consequences on water quality (**Figure 1.1**).

There are many cases in which water quality declines are due to changed agricultural activities, as discussed by Bates et al. (2008) and others. The cases include, but are not limited to, the following.

- Nutrient loads from agricultural land during more extreme storms (Bouraoui et al., 2002).

- Less diluted nutrients due to reduced flows in summer, particularly with drought events (Whitehead et al., 2006).

- Irrigation return flow can affect water quality in areas with increased irrigation water uses through traditional irrigation systems such as flooding systems.

- An increasing use of fertilisers and pesticides due to land-use change and longer growing seasons, with subsequent leaching to water compartments (Moss et al., 2004; Bloomfield et al., 2006).

- The development of biofuels, which can have a complex impact on water quality. Expanding demand for biofuels creates incentives to increase production, and thus on the use of inputs such as fertilisers and pesticides. Cellulosic feedstocks such as Miscanthus consume much higher amounts of water during their growth period and can reduce drainage and runoff, particularly low flows, which in turn may cause water quality problems due to less dilution (McIsaac et al., 2010). However, cellulosic-based feedstocks can offer considerable potential to reduce adverse impacts on feedstock production in terms of water quality because these crops need very little applied nitrogen (NRC, 2007; Ng et al., 2010).

Studies on water quality under climate change at the regional scale examine, illustrate or predict the impacts listed above by conducting historical data analysis and using simulation modelling tools. For example, the southwestern **United States** will suffer from water stress and severe water quality problems in agricultural regions (Cruise et al., 1999). The water flow and nutrient concentrations will increase (Arnell, 1998; Bouraoui et al., 2002) in the **United Kingdom**, whereas the stream flow and nutrient load will decrease in Greece (Varanou et al., 2002). **Finland** will experience fewer changes in annual water runoff but a clear difference in seasonal patterns and higher nutrient loads (Kallio et al., 1997). In Scotland, the direct impacts of climate change on hydrological functioning and nitrate pollution would be less than those caused by the land use change; the changes of nitrate pollution may depend on location, season and the climate scenario (Dunn et al., 2012). Watersheds in western **Japan** could experience trophic lake conditions with risk of eutrophication (Komatsu et al., 2007).

A recent study by Jeppesen et al. (2011) found that the projected climate change will most probably enhance the nitrogen and phosphorus load of lakes in northern **Europe**, especially in winter. In arid southern Europe, although nutrient loads may decrease, nitrogen and phosphorous concentrations may increase due to higher evaporation, which in turn will lead to a decrease in water quality and ecological status. To counteract this deterioration, more sustainable agriculture with less loss of nutrients to surface waters is recommended for the North temperate zone, including: better management of animal manure and chemical fertilisers, implementation of crop rotations, improved exploitation of feed-stuffs, and reduction of nitrogen application

Climate change and extreme water events: Droughts and floods

Recent experiences with extreme events have led to a growing interest in how extremes affect agricultural outcomes. There is growing evidence that climate change is likely to increase the frequency and severity of extreme weather events (IPCC, 2012). Increased evapotranspiration would lead to greater overall levels of precipitation, but precipitation events may become heavier and less frequent, and the geographical patterns of rainfall may change. With more heat, evaporation and intermittent rainfall, droughts and floods may thus become more frequent and severe. The objective of this section is to review the current state of knowledge on the impact of climate change on extreme weather events related to the water cycle, and their related consequences on the agricultural sector.

An *extreme climate event* – also referred to as extreme weather event or climate extreme – is defined as the "occurrence of a value of a weather (...) variable above (or below) a *threshold* value near the upper (or lower) ends of the range of observed values of the variable" (IPCC, 2012). Once this threshold is defined, the *probability* of the considered extreme event depends on the shape of the statistical distribution of the weather variable (**Figure 1.7**, which displays three hypothetical examples for the case of temperature). On each graph, the probability of extreme hot weather event is equal to the dark grey area on the right side of the statistical distribution, while the probability of an extreme cold weather is equal to the dark blue area on the left side. Looking at these three different cases (graphs a, b and c), one can see that a change in the probability of an extreme event, whether hot or cold, can result from the following.

- A shift in the whole statistical distribution on the right, resulting in a change in the mean (Graph A of **Figure 1.7**).

- An increase in the variability of the statistical distribution without a change in the mean (Graph B).

- A change in symmetry of the statistical distribution (Graph C).

These hypothetical examples show it is important to consider both sides of the statistical distribution when dealing with extreme weather events, and not just focus, for example, on extreme hot weather events. Indeed, a shift in the whole statistical distribution of temperature on the right implies an *increase* in the probability of extreme hot weather events but a *decrease* in extreme cold weather events. In contrast, an increase in the variability of temperature would lead to an *increase in both extremes* – hot and cold weather events. These two situations may result in significantly different impacts on agricultural production systems.

Other important issues are the *temporal and spatial scales* of the weather variables considered. In the example of temperature, one has to choose between a large set of candidate variables such as: average daytime temperature, average night temperature, maximum day temperature, average growing season temperature, etc. Temperature can also be aggregated at the spatial scale, such as the region, county, etc., which does not perfectly reflect local climate conditions. As recently underlined by Ortiz-Orbea and Just (2012), the choice of methodological approaches and weather variables can significantly affect the assessment of the impacts of extreme weather events – and more generally climate change – on agricultural production.

It is also important to consider the case of *compound events*, which relates to situations where the combination of several weather variables can amplify the extreme nature of the event – or its impact – or even create it. In cases of compound events, the thresholds that define extreme weather events variable by variable are no longer sufficient when considered in isolation. For example, farmers may face for the same growing season extreme winter rainfall and then a drought during the summer, leading to substantial crop losses. More subtle would be a case of high – but not extreme – winter rainfall followed by hot – but not extreme – summer temperature. The combination of these two "not extreme" events could eventually lead to substantial crop losses as well. In this case, each event is not statistically rare considered in isolation, but the occurrence of the two in the same growing season can be. To be consistent with the IPCC definition of extreme weather event, we should thus consider the *joint probability distribution* of the two weather variables, instead of looking at each one separately.

Figure 1.7. The effect of changes in temperature distribution on extremes

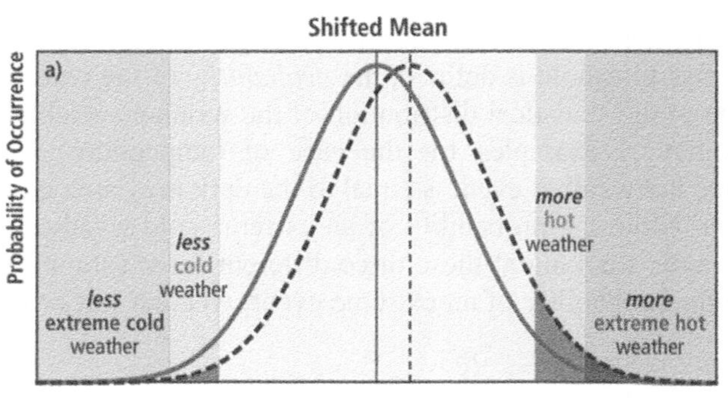

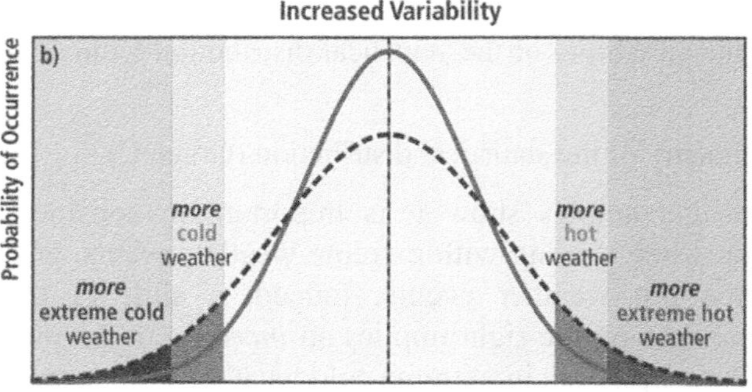

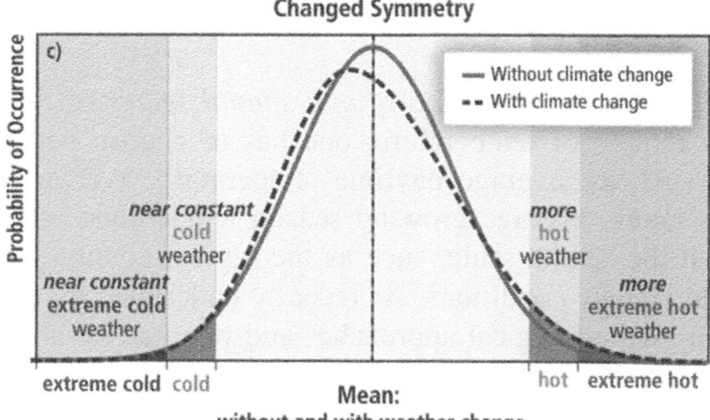

Notes: Different changes in temperature distributions between present and future climate and their effects on extreme values of the distributions:
(a) Effects of a simple shift of the entire distribution toward a warmer climate;
(b) Effects of an increase in temperature variability with no shift in the mean;
(c) Effects of an altered shape of the distribution, in this example a change in asymmetry toward the hotter part of the distribution.

Note that the Gaussian shapes of the statistical distributions presented here are hypothetical, for the purpose of illustration. In reality, statistical distributions of weather variables such as temperature and rainfall can take more specific forms.

Source: IPCC (2012), *Managing the Risks of Extreme Events and Disasters to Advance Climate Change Adaptation. A Special Report of Working Groups I and II of the Intergovernmental Panel on Climate Change*, Cambridge University Press, Cambridge, UK, and New York, NY, US.

Why not simply focus on average projections? Although extremes are likely to have the strongest influence on agricultural outcomes, analysis that successfully measures *average effects* of climate might provide reasonable first-order approximation of impacts from climate change. This is, however, not always the case and depends on the shape of the relationship between the weather variable (temperature, precipitation, etc.) and its impact on agriculture. More precisely, one can state that focusing on average effects can be a reasonable approximation if the following conditions are fulfilled:

- either the relationship between the effect of climate change and the impact is *linear*; or

- if the relationship is nonlinear and the change in climate and its effects are *smooth, gradual* and *relatively small.*

In contrast, analysis of average effects will not suffice if the effect of climate change on the outcome is nonlinear or discontinuous and the change in climate is large, or weather becomes more variable with climate change. **Box 1.3** explains more precisely this issue.

Box 1.3. Non-linear effects and the cost of risk

The issue of non-linearity arises from the fact that with a non-linear relationship between yield and random weather variable, the average outcome is not equal to the outcome of the average value of the weather variable. Suppose that the yield of a given crop depends on a weather variable such as temperature or rainfall, and that crop yield is an increasing and concave function of the weather variable, all things being equal. Suppose that there is equal probability that a weather variable equals either a low value W_L or a high value W_H. The average value of the weather variable (such as precipitation) thus equals $(W_L + W_H)/2$. Using this average value of the weather variable to predict yield would thus lead to a value of $Y((W_L + W_H)/2)$, which is represented by point A. This value of yield is higher than the expected yield computed on the distribution of the weather variable, which is equal in our hypothetical example to $(1/2)*Y(W_L) + (1/2)*Y(W_H)$ and corresponds to the horizontal dashed line in **Figure 1.8**. This illustrates graphically a more general statement that using average values of weather variables instead of statistical distributions for predicting the impacts of climate change on agricultural outcomes is likely to result in a systematic bias. The empirical challenges are to discern whether the climate changes we have had and are likely to experience over the coming decades can be considered small enough for a linear approximation to suffice, and climate changes acts primarily in manner that shifts the average weather outcomes and not variability around the average outcome.

Figure 1.8. Non-linear impacts of weather variables on agricultural outcomes

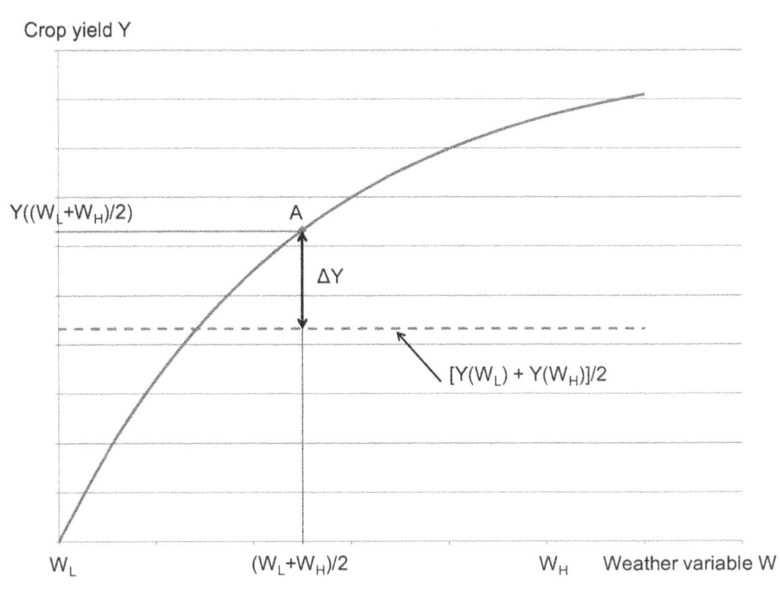

Droughts: Past trends and projected impacts on agriculture under climate change

Droughts come with a combination of low precipitation and high temperatures, which affect water quantity and quality and can drastically decrease agricultural production (Calanca, 2007; Mishra and Singh, 2010; Yao et al., 2011; Eitzinger et al., 2012; Fraser et al., 2012). While it seems that at the global scale it is difficult to see any change in frequency of drought (Sheffield et al., 2012), many regions in the world have faced more frequent and severe droughts over the past decades leading to significant damages (Mishra and Singh, 2010). Droughts are expected to be more intense and likely to happen more frequently due to climate change in the Alpine region, the United States, the Mediterranean basin, Australia, and South Africa and many other places around the world (Calanca, 2007; Planton et al., 2008; Wang et al., 2011; Mpelasoka et al., 2008). Drier soils and more frequent droughts are expected in the June–August season in Amazon and West Africa, and in the December–February season in the Asian monsoon region (Wang, 2005; Fraser et al., 2012).

Heat waves are usually related to drought events and refer to persistent elevated temperatures, causing eventual damages to crop and livestock production (Beniston and Diaz, 2004). Changes in frequency and intensity of heat waves have already been observed in Europe (Klein Tank and Können, 2003), in China (Zhou et al., 2012; Jiang et al., 2012), in India (Dash and Mamgain, 2011), in Africa (Aguilar et al., 2009), and in the rest of the world (IPCC, 2012, Annex B). Heat waves are expected to become more frequent and severe in many places due to climate change (Meehl and Tebaldi, 2004; Albright et al., 2011). The combination of drought and heat waves intensifies and broadens the drought impacts on agricultural production (Calanca, 2007).

Figure 1.9. Non-linear effects of temperature on crop yields

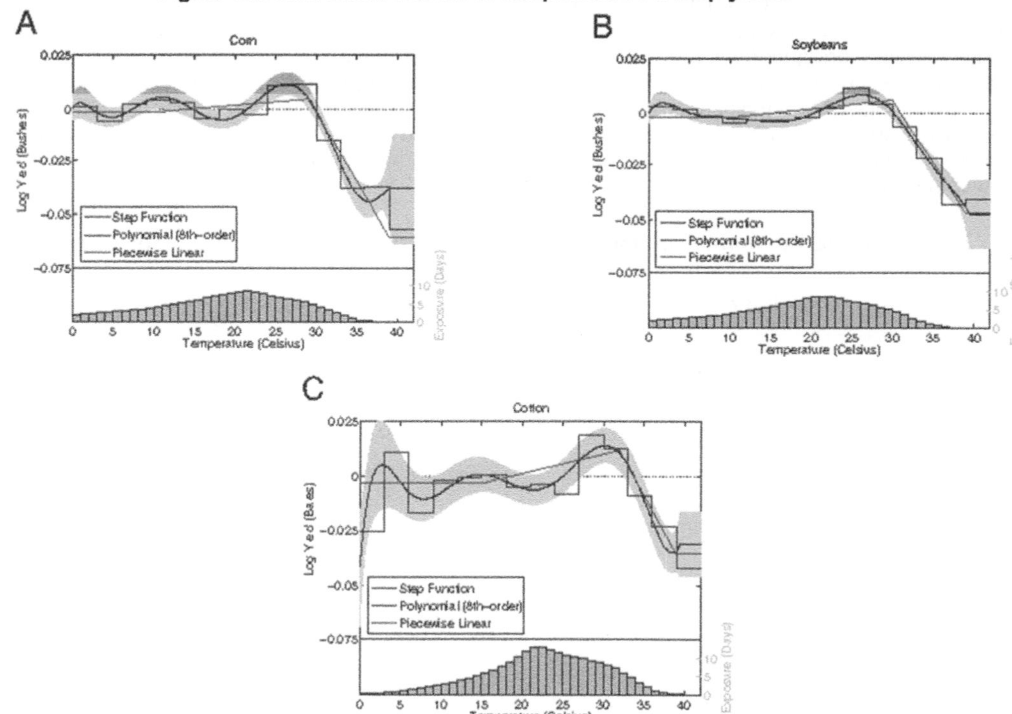

Source: Schlenker, W. and M. Roberts (2009), "Nonlinear temperature effects indicate severe damages to US crop yields under climate change", *Proceedings of the National Academy of Sciences*, Vol. 106(37).

More frequent and severe droughts due to climate change in the future could have substantial and negative impacts on agricultural production, due to the sensitivity of crop yields on temperature beyond certain thresholds. Recent research by Schlenker and Roberts

(2009) underlines the highly non-linear impact of temperature on crop yields in the **United States**: there does appear to be a temperature threshold above which crop outcomes quickly decline. For corn, soybeans and cotton, estimated critical temperature thresholds are of 29 C, 30∘C and 32∘C, respectively (**Figure 1.9**). Declines in yield for temperatures above these thresholds were far larger than declines below the threshold. A number of studies have found similar predictions using a variety of temperature and rainfall measures. Lobell et al. (2011b) shows similar sensitivity to extreme heat across field experiment stations in Africa for corn. Research by Lobell and Field (2007) and Lobell et al. (2011a) summarises global-scale evidence from the largest staple crops.

In the **European Union**, the 2003 heat wave caused the warmest summer since 1540, with increased evaporative requirement, increased irrigation and decreased crop yield (Jolly et al., 2005; van der Velde et al., 2010; Garcia-Herrera et al., 2010). This event caused a 30% reduction of the gross primary productivity (Ciais et al., 2005). Beniston (2004) recommends this event to be used as a climatic analog by scientists and decision makers for developing adaptation strategies as similar heat waves are expected to occur more frequently based on the many projections for the end of the 21st century in the region (Meehl and Tebaldi, 2004). **Figure 1.10** shows changes in the recurrence of 100-year droughts based on comparisons between climate and water use in 1961-1990 and simulations for the 2020s and 2070s (Lehner et al., 2006).

Figure 1.10. Changes in the recurrence of 100-year drought based on comparisons between climate and water use in 1961-1990 and simulations for the 2020s and 2070s

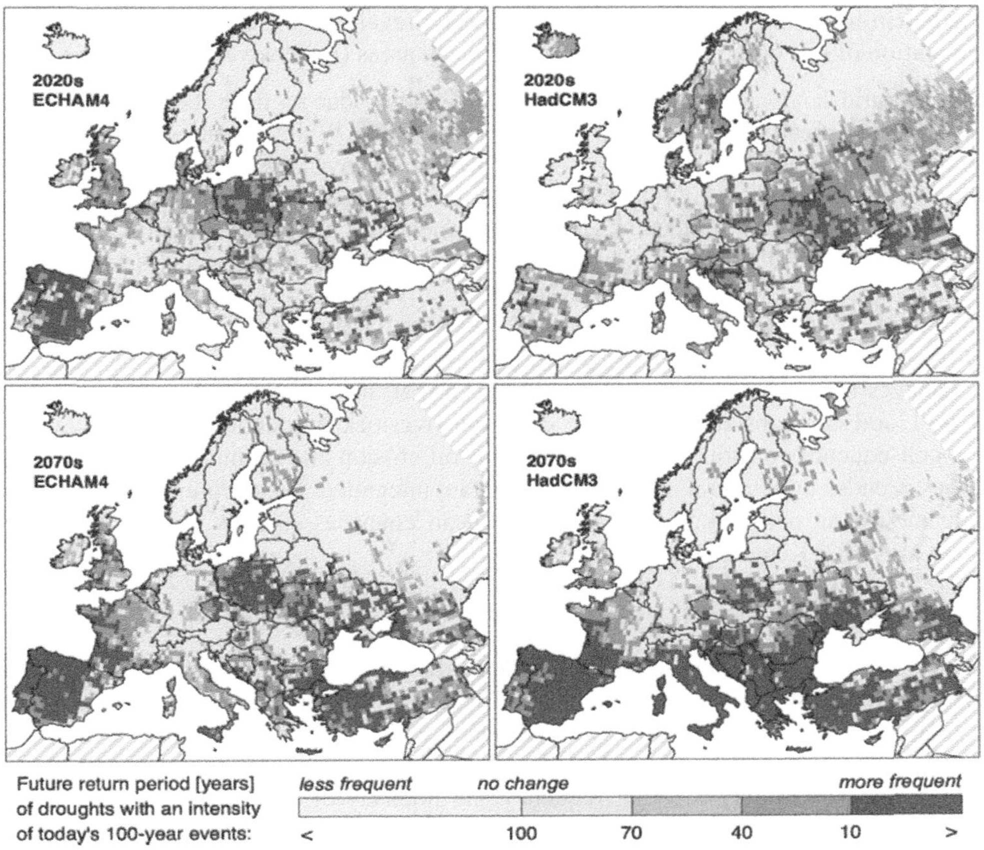

Source: Lehner, B., Döll, P., Alcamo, J. Henrichs, T. and F. Kaspar (2006), "Estimating the impact of global change on flood and drought risks in Europe: a continental, integrated analysis", *Climatic Change*, Vol. 75.

There are several challenges in assessing the impact of droughts related to climate change. Droughts cause losses in agriculture through multiple factors including water deficit and heat, but also more indirect ones such as pests and diseases. For example, in Central and Eastern Europe, crop productivity usually drops due to spreading of insects during droughts (Eitzinger et al., 2012). Droughts could affect crop production by reducing water reserves available for irrigation. For instance, winter heat waves in Europe could lead to earlier than usual snow melting and consequently affect water resources management strategies (Beniston, 2007). Constant increases in winter temperature, as suggested by most climate change projections, together with more frequent heat waves in winter, may lead to decreases in snow pack storage, a potential source of water supply for irrigation in semi-arid regions.

Floods: Past trends and projected impacts on agriculture under climate change

Changes in the frequency and intensity of precipitation extremes will also probably occur over the next decades, which may cause increasing agricultural production losses (Tubiello et al., 2007). At present, the impact has already been considerable. For example, excess soil moisture resulting from heavy rainfalls has been the main cause for both insurance indemnities and disaster payments for agriculture in California (Lobell et al., 2011c). US corn production loss due to excess soil moisture is expected to double by 2030 under climate change (Rosenzweig et al., 2002). Excess precipitation also causes agricultural losses due to pests and plant diseases, as observed in the Netherlands and the United States (Schaap et al., 2011; Rosenzweig et al., 2001), delayed field work (Rosenzweig et al., 2001), soil erosion especially in Europe and the Mediterranean basin (Grimm et al., 2002; Fuhrer et al., 2006), as well as environmental problems such as epidemics, prevalence of leaf fungal pathogens, and dissemination of soil borne pathogens to non-infected areas (Rosenzweig et al., 2001).

Agricultural lands are especially exposed to flooding due to their location in or nearby flood expansion areas and floodplains (Förster et al., 2008). While the cases of flood losses in urban areas are well analysed (Smith, 1994; Browne and Hoyt, 2000; Burby, 2001), the study of flood losses in the agricultural sector has not gained much attention (Förster et al., 2008; Tapia-Silva et al., 2011). Besides *in situ* data and modeling tools (Dutta et al., 2003), remote sensing methods have the potential to assess flood losses in agriculture (Pantaleoni et al., 2007; Tapia-Silva et al., 2011). However, the relative economic losses are quite limited compared to urban areas that concentrate substantially higher capital per unit of land. According to Messner et al. (2007), agricultural losses are estimated between 1% and 5% of total damages of flood events. This does not mean that the impacts of floods are limited for farmers. Flood can substantially reduce crop and livestock production, and can have long-lasting consequences on soil productivity due to soil erosion and draining problems. In terms of policy trends, agricultural lands as part of an integrated flood control management is gaining increasing consideration in some European countries such as France and Scotland (Pivot et al., 2002; Kenyon et al., 2008).

Other extreme weather events

Hurricanes, cyclones and typhoons – depending on the geographical location – are perhaps the most destructive extreme climate events. They combine very strong winds and very high precipitation and are often associated with sea level rise, and each of these phenomena taken separately is susceptible of wreaking havoc on agricultural land. Under climate change, it is likely that the frequency and intensity of hurricanes will change in the future, but the uncertainty on the trends remains high, especially at the global level (Meehl et al., 2000; Knutson and Tuleya, 2004). However, most studies have focused on infrastructure damages and human casualties rather than agriculture, which appear secondary. In Viet Nam, economic losses caused by typhoons seem to have increased from the 1950s to the 1990s despite a noticeable decrease in the frequency of these events (Imamura and

Van To, 1997). In Mexico, coffee-growing regions are threatened by hurricanes and the landslides caused by these events (Philpott et al., 2008). In the United States, hurricanes impact crop acreage and crop prices in positive or negative ways at the national level, depending if the region is stricken or non-stricken; crop acreages tend to move from stricken to non-stricken regions and welfare also moves from stricken to non-stricken regions (Chen and McCarl, 2009).

The opposite temperature extreme is the cold spell, i.e. abrupt temperature drops. Particular types of cold spell include frost and spring frost (also called freeze and spring freeze), which can potentially cause substantial damage to crops. Because of the increase of the mean temperature at the global scale, the number of cold days occurring each year have been decreasing and are expected to keep decreasing in the future (IPCC, 2007b; Park et al., 2011). However, in a world with a changing climate, spring frost risk is bound to change because it is sensitive to variations in daily temperature variance and mean temperature (Rigby and Porporato, 2008). The 2007 eastern US spring freeze highlighted the possibility of increasing risk of spring freeze due to climate change (Gu et al., 2008). In particular, because of higher temperatures, crops can develop prematurely and therefore are more vulnerable to spring freeze (Gu et al., 2008; Marino et al., 2011).

Remarks on extreme climate events

Changes in temperature and precipitation extremes have been observed and are expected in the future, for example – but not limited to – in China (Zhou et al., 2012; Jiang et al., 2012), in Europe, in India (Dash and Mamgain, 2011), in Africa (Aguilar et al., 2009) and the rest of the world (IPCC, 2012). High uncertainty remains with how these changes will impact agriculture, either in beneficial or adverse ways, depending on the trend – decrease or increase in the frequency and magnitude of the extreme events – and the region considered (Gao and Zhao, 2002; Shabbar and Bonsal, 2003). Extreme climate events vary highly with time and space, which makes them difficult to predict. It seems that extreme events occur more frequently in some regions while as frequently or even less frequently in others (Imamura and Van To, 1997, Sheffield et al., 2012). Indeed, occurrence of typhoons is reported to have decreased since the 1950s in Viet Nam (Imamura and Van To, 1997). Frequency of cold spells has decreased in China while the frequency of warm spells has increased and frequency of heavy rainfalls has also increased in some regions of China (Gao and Zhao, 2002). The spatial variability of both adverse and beneficial effects of climate change is also illustrated in Canada, where particularly cold spells have decreased in frequency, duration and intensity in some regions while increased in other regions; winter warm spells have increased in frequency and duration in most parts of the country but have decreased in northeastern regions (Shabbar and Bonsal, 2003). Furthermore, the frequency of droughts has seen little change at the global scale during the past 60 years (Sheffield et al., 2012). However, even if these events do not occur frequently, the potential of extreme weather events to wreak havoc on agricultural lands puts them in the high risk category regarding agriculture.

Although the literature review reveals the fact that damages and losses are significant, more work is needed to understand the impacts of extreme climate events on agriculture. The impacts could be counterbalanced by the direct effects of climate change on agriculture, such as increased levels of CO_2 in the atmosphere (Rosenzweig and Parry, 1994; Mendelsohn et al., 1994; Parry et al., 2004), increased precipitation (Mendelsohn et al., 1994), increased temperatures, and by human activities, such as crop bioengineering aiming to create drought or flood resistant crops (Mitra, 2001). Each significant extreme event that has occurred in the past, such as the 2003 heat wave in Europe and 2007 freeze in the United States, should be assessed clearly and its impacts on agriculture, its future frequency and magnitude, and possible actions to undertake should be studied thoroughly. The evolution of extreme event impacts under climate change is perhaps the most uncertain question as of now, and remains a crucial subject that requires more attention from both scientists and policy makers.

Summary

Before moving to policy analysis of adaptation strategies for agricultural water management, this section summarises the main outcomes on the impacts of climate change on the water cycle and agriculture.

Generally the projections indicate that climate change could lead to an acceleration of the water cycle, involving more frequent and intense rainfall episodes, however, there are differential impacts in different regions so that projections indicate increases in precipitation at high latitudes in winter and summer seasons, and decreases in summer precipitation in mid-latitude and sub-tropical area.

In addition to water availability also crop water requirements will change, which will have implications for irrigation water demand. Irrigation water demand is likely to decline slightly at the global scale despite the projected warmer climate, although changes vary by region. Lower demand can be expected in Africa, Australia and China, while for other regions —including Europe, North India, eastern portion of South America and the eastern United States—uncertainty as regards projections is large.

As regards crop yields it is likely that water could become a major limiting factor for crop production in the context of climate change. Climate variability and extreme climatic conditions will also affect livestock growth since climate conditions (e.g. temperature, humidity, wind speed and precipitation) affect animal performances, including weight, milk production, wool production, and reproduction directly or indirectly due to the quality and quantity of feedstuffs and the severity and distribution of livestock diseases.

As regards water quality climate change affects agricultural land use and production practices that in turn have impacts on water quality. The mechanisms at stake are, however, complex and involve chain reactions due to adaptation of agricultural management practices that are difficult to predict. The future importance of pests, diseases and weeds remains a quite underexplored area of projections, and so the future trends in pesticide use.

There is growing confidence that climate change will increase the frequency and severity of extreme weather events such as droughts and floods. Moreover, this is not just an increase in risk, but also an increase in uncertainty, which means that the set of events and their associated probabilities will be continuously evolving due to the non-stationary nature of the changing climate. This increase in uncertainty is perhaps a greater challenge than the increase in risk itself. Not only is there uncertainty about shifts in extremes, there is also low predictability since most climate models don't even try to model shifts in extremes but only extrapolate changes in extremes from shifts in means. Still, it seems that extreme events occur more frequently in some regions, and as frequently or even less frequently in others.

The evidence base for taking decisions about projected climate impacts on water may be quite weak, since climate projections are typically very uncertain for precipitation and thus downscaling is problematic. Relative to average temperature change projections of precipitation patterns remain highly uncertain, in particularly at local level. This means that projections at the resolution and scale required for farmers' adaptation decisions are lacking. Hence, typical assessments of the impacts of climate change on water may be of limited use when it comes to making practical, on-site decisions about adaptation. More generally, the level of confidence in climate change projections decreases as their potential utility for making decisions on how to adapt increases. Consequently, adaptation decisions need to accommodate considerable uncertainty.

There are other major drivers than climate change that affect water use in agriculture, and are perhaps more certain than climate change projections: increase in the world population and changes in dietary habits, which are expected to increase the demand for food, and rising competition between water uses. These major drivers are strongly expected to put substantial additional pressure on water systems, on both quantity and quality dimensions. To the climate trends should thus be added these socio-economic trends to get a full picture of the impacts of climate change on the water cycle and agriculture.

After having presented the main impacts of climate change on water systems and agriculture, the issue is to find adaptation strategies that are in line with these projections, but are also able to deal with the inherent and unsurpassable levels of uncertainty that will remain in decision making. The next chapter on adaptation strategies aims at proposing a framework and analysing the main policy approaches for addressing adaptation of agricultural water management to climate change in such a complex environment.

References

Aguilar, E., A. Aziz Barry, M. Brunet, L. Ekang, A. Fernandes, M. Massoukina, J. Mbah, A. Mhanda, D.J. do Nascimento, T.C. Peterson, O. Thamba Umba, M. Tomou and X. Zhang (2009), "Changes in temperature and precipitation extremes in western central Africa, Guinea Conakry, and Zimbabwe, 1955–2006", *Journal of Geophysical Research*, Vol. 114, No. D2.

AIACC (2006a), *Climate change and variability in the mixed crop-livestock production systems of the Argentinean, Brazilian and Uruguayan Pampas*, AIACC, Project No. LA 27.

AIACC (2006b), Climate Change Vulnerability and Adaptation in the Livestock Sector of Mongolia, AIACC, Project No. AS 06.

Albright, T.P., A. M. Pidgeon, C.D. Rittenhouse, M.K. Clayton, C.H. Flather, P.D. Culbert, and V.C. Radeloff (2011), "Heat waves measured with MODIS land surface temperature data predict changes in avian community structure", *Remote Sensing of Environment*, Vol. 115, No. 1, pp. 245-254.

Anwar, M.R., G. O'Leary, D. McNeil, H. Hossain and R.Nelson (2007), "Climate change impact on rainfed wheat in south-eastern Australia", *Field Crops Research* 104, pp. 139-147.

Arnell, N.W. (1998), "Climate change and water resources in Britain", *Climatic Change*, Vol. 39, No. 1, pp. 83–110.

Bates, B.C., Z.W. Kundzewicz, S. Wu and J.P. Palutikof (eds.), (2008), *Climate Change and Water. Technical Paper of the Intergovernmental Panel on Climate Change*, IPCC Secretariat, Geneva, 210 pp.

Beniston, M. (2007), "Linking extreme climate events and economic impacts Examples from the Swiss Alps", *Energy Policy*, Vol. 35, pp. 5384-5392.

Beniston, M. (2004), "The 2003 heat wave in Europe: A shape of things to come? An analysis based on Swiss climatological data and model simulations", *Geophysical Research Letter*, Vol. 31, p. 02202.

Beniston, M. and H.F. Diaz (2004), "The 2003 heat wave as an example of summers in a greenhouse climate. Observations and climate model simulations for Basel, Switzerland", *Global and Planetary Change*, Vol. 44, pp. 73-81.

Bloomfield, J.P., R.J. Williams, D.C. Goody, J.N. Cape and P. Guha (2006), "Impacts of climate change on the fate and behaviour of pesticides in surface and groundwater – a UK perspective", *Science of the Total Environment*, Vol. 369, No. 1-3, pp. 163–177.

Bouraoui, F., L. Galbiati and G. Bidoglio (2002), "Climate change impacts on nutrient loads in the Yorkshire Ouse catchment", *Hydrology and Earth Systems Sciences Discussions*, Vol. 6, No. 2, pp. 197–209.

Browne, M.J., and R.E. Hoyt (2000), "The demand for flood insurance: empirical evidence", *Journal of risk and uncertainty*, Vol. 20, No. 3, pp. 291-306.

Burby, R.J. (2001), "Flood insurance and floodplain management: the US experience", *Environmental hazards*, Vol. 3, pp. 111-122.

Cai, X., D. Wang and R. Laurent (2009), "Impact of climate change on crop yield: a case study of rainfed corn in central Illinois", *Journal of Applied Meteorology Climatology*, Vol. 48, pp. 1868-1881.

Cai, X., X. Zhang, P. Noël and M. Shafiee-Jood (2013), "Impact of Climate Change on Water Quantity and Quality and Implications to Agriculture – A Review", unpublished consultant report.

Calanca, P. (2007), "Climate change and drought occurrence in the Alpine region. How severe are becoming the extremes", *Global and Planetary Change*, Vol. 57, pp. 151-160.

Chen, C. and B. McCarl (2009), "Hurricanes and possible intensity increases: effects on and reactions from US agriculture", *Journal of Agricultural and Applied Economics*, Vol. 41, p. 125.

Ciais, PH. et al. (2005), "Europe-wide reduction in primary productivity caused by the heat and drought in 2003", *Nature*, Vol. 437, No. 22.

Crimp, S., M. Howden, B. Power, E. Wang, and P. De Vo, (2008), "Global Climate Change impacts on Australia's Wheat Crops," Report prepared for the Garnaut Climate Change Review Secretariat, March 2008, p. 16.

Cruise, J.F., A.S. Limaye and N. Al-Abed (1999), "Assessment of impacts of climate change on water quality in the southeastern United States", *Journal of the American Water Resources Association*, Vol. 35, No. 6, pp. 1539–1550.

Dash, S.K. and A. Mamgain (2011), "Changes in the Frequency of Different Categories of Temperature Extremes in India", *Journal of applied meteorology and climatology*, Vol. 50, pp. 1842–1858.

Delpla, I., A.-V. Jung, E. Baures, M. Clement and O. Thomas (2009), "Impacts of climate change on surface water quality in relation to drinking water production", *Environment International*, Vol. 35, pp. 1225–1233.

Dillon, P.J., L.A. Molot and M. Futter (1997), "The effect of El Nino-related drought on the recovery of acidified lakes", *Environmental Monitoring and Assessment*, Vol. 46, pp. 105–111.

Döll, P. (2002), "Impact of Climate Change and Variability on Irrigation Requirements: A Global Perspective", *Climatic Change*, Vol. 54, pp. 269-293.

Dunn, S.M., I. Brown, J. Sample and H. Post (2012), "Relationships between climate, water resources, land use and diffuse pollution and the significance of uncertainty in climate change", *Journal of Hydrology*, 434–435, pp. 19-35.

Dutta, D., S. Herath and K. Musiake (2003), "A mathematical model for flood loss estimation", *Journal of Hyrdology*, Vol. 277, pp. 24-49.

EEA (2012), "Climate change, impacts and vulnerability in Europe 2012 – An indicator-based report". European Environment Agency.

Eitzinger, J., D. Trnka, S. Semeradova, S. Thaler, E. Svobodova, P. Hlavinka, B. Siska, J. Takac, L. Malatinska, M. Novakova, M. Dubrovsky and Z. Zalud (2012), "Regional climate change impacts on agricultural crop production in central and eastern Europe-hotspots, regional differences and common trends", *Journal of Agricultural Science*, pp. 1-26, doi:10.1017/S0021859612000767.

FAO (2011), *Climate change, Water and Food Security*, FAO Water Report No. 36, Food and Agriculture Organization of the United Nations (FAO), Rome.

FAO (2008), Crop evapotranspiration – Guidelines for computing crop water requirements – FAO Irrigation and drainage paper 56.

Fellmann, T. (2012), "The assessment of climate change-related vulnerability in the agricultural sector: reviewing conceptual frameworks", in Meybeck, A., Lankoski, J., Redfern, S., Azzu, N. and V. Gitz (eds.) *Building resilience for adaptation to climate change in the agriculture sector – Proceedings of a Joint FAO/OECD Workshop*, Rome, 23-24 April, 2012.

Fischer, G., F.N. Tubiello, H. van Velthuizen and D.A. Wiberg (2007), *Climate change impacts on irrigation water requirements: Effects of mitigation, 1990–2080*, Technological Forecasting and Social Change Vol. 74, pp. 1083-1107.

Förster, S., B. Kuhlmann, K.E. Lindenschmidt and A. Bronstert (2008), "Assessing flood risk for a rural detention area", *Natural Hazards and Earth System Science*, Vol. 8, pp. 311–322.

Frank, K.L., T.L. Mader, J.A. Harrington, G.L. Hahn and M.S. Davis (2001), "Climate Change Effects on Livestock Production in the Great Plains", pp. 351-358 in *Livestock Environment VI: Proceedings of the 6th International Symposium* (21-23 May 2001, Louisville, Kentucky, United States).

Fraser, E.D.G., E. Simelton, M. Termansen, S.N. Gosling and A. South (2013), "Vulnerability hotspots: Integrating socio-economic and hydrological models to identify where cereal production may decline in the future due climate change induced drought", *Agricultural and Forest Meteorology*, Volume 170, pp. 195–205.

Fuhrer, J., M. Beniston, A. Fischlin, C.H. Frei, S. Goyette, K. Jasper and CH. Pfister (2006), "Climate risks and their impacts on agriculture and forests in Switzerland", *Climatic Change*, Vol. 79, pp. 79–102.

Gao, X. and Z. Zhao (2002), "Changes of Extreme Events in Regional Climate Simulations over East Asia", *Advances in Atmospheric Sciences*, Vol. 19, No. 5.

García-Herrera, R., J. Díaz, R.M. Trigo, J. Luterbacher and E.M. Fischer (2010), "A Review of the European Summer Heat Wave of 2003", *Critical Reviews in Environmental Science and Technology*, Vol. 40, No. 4, pp. 267-306.

Grimm, M., R. Jones and L. Montanarella (2002), *Soil Erosion Risk in Europe*, European Soil Bureau Institute for Environment & Sustainability JRC Ispra. EUR 19939 EN © European Communities.

Gu, L., P.J. Hanson, W. Mac Post, D.P. Kaiser, B. Yang, R. Nemani, S.G. Pallardy and T. Meyers (2008), "The 2007 Eastern US Spring Freeze: Increased Cold Damage in a Warming World?", *BioScience*, Vol. 58, No. 3, pp. 253-262.

Guiteras, R. (2007), "The impact of climate change on Indian agriculture", Department of Economics, MIT.

Harle, K.J., S.M. Howden, L.P. Hunt and M. Dunlop (2007), "The potential impact of climate change on the Australian wool industry by 2030", *Agricultural Systems*, Vol. 93, pp. 61-89.

Henry, B., E. Charmley, R. Eckard, J.B. Gaughan, and R. Hegarty (2012), "Livestock production in a changing climate: adaptation and mitigation research in Australia", *Crop and Pasture Science*, Vol. 63, pp. 191-202.

Holden, N.M. and A.J. Brereton (2003), "Potential impacts of climate change on maize production and the introduction of soybean in Ireland", *Irish Journal of Agriculture and Food Research*, Vol. 42, pp. 1-15.

Howden, M. and R. Jones (2004), "Risk assessment of climate change impacts on Australia's wheat industry", In *Proceedings for the 4th International Crop Science Congress*, Brisbane, Australia, 26 September-1 October. Available at: www.cropscience.org.au.

Howden, S.M., S.J. Crimp and C.J. Stokes (2008), "Climate change and Australian livestock system: research, impacts and policy issues", *Australian Journal of Experimental Agriculture*, Vol. 48, pp. 780-788.

IFPRI (2009), "Climate Change – Impact on Agriculture and Costs of Adaptation", Food Policy Report International Food Policy Research Institute, Washington, DC, USA, p. 30.

Iglesias, A., Quiroga, S. and A. Diz (2011), 'Looking into the future of agriculture in a changing climate', *European Review of Agricultural Economics*, Vol. 38(3), pp. 427-447.

Imamura, F. and D. Van To (1997), "Flood and Typhoon Disasters in Viet Nam in the Half Century Since 1950", *Natural Hazards*, Vol. 15, No. 1, pp. 71–87.

IPCC (2013), *Climate Change 2013, The Physical Science Basis – Summary for Policy Makers*, Working Group I Contribution to the Fifth Assessment Report of the Intergovernmental Panel on Climate Change.

IPCC (2012), *Managing the Risks of Extreme Events and Disasters to Advance Climate Change Adaptation*. A Special Report of Working Groups I and II of the Intergovernmental Panel on Climate Change, Cambridge University Press, Cambridge, UK, and New York, NY, USA, 582 pp.

IPCC (2007a), *Climate Change 2007: Impacts, Adaptation and Vulnerability. Contribution of Working Group II to the Fourth Assessment Report of the Intergovernmental Panel on Climate Change*, M.L. Parry, O.F. Canziani, J.P. Palutikof, P.J. van der Linden and C.E. Hanson, Eds., Cambridge University Press, Cambridge, UK, pp. 976.

IPCC (2007b), *IPCC 2007: Climate Change 2007: The Physical Science Basis. Contribution of Working Group I to the Fourth Assessment*, Report of the Intergovernmental Panel on Climate Change, Cambridge University Press, Cambridge, United Kingdom and New York, NY, USA, pp. 996.

Jeppesen, E., B. Kronvang, J. Olesen, J. Audet, M. Søndergaard, C. Hoffmann, H. Andersen, T. Lauridsen, L. Liboriussen, S. Larsen, M. Beklioglu, M. Meerhoff, A. Ozen and K. Ozkan (2011), "Climate change effects on nitrogen loading from cultivated catchments in Europe: implications for nitrogen retention, ecological state of lakes and adaptation". *Hydrobiologia*, 663, pp. 1-21.

Jiang, D., Z. Li and Q. Wang (2012), "Trends in temperature and precipitation extremes over Circum-Bohai-Sea region, China", *Chinese Geographical Science*, Vol. 22, No. 1, pp. 75–87.

Jolly, W.M., M. Dobbertin, N.E. Zimmermann and M. Reichstein (2005), "Divergent vegetation growth responses to the 2003 heat wave in the Swiss Alps", *Geophysical Research Letters*, Vol. 32, No. 18.

Kallio, K., S. Rekolainen, P. Ekholm, K. Granlund, Y. Laine, H. Johnsson and M. Hoffman (1997), "Impacts of climatic change on agricultural nutrient losses in Finland", *Boreal Environment Research* Vol. 2, pp. 33–52.

Kabubo-Mariara, J. (2009), "Global warming and livestock husbandry in Kenya: Impacts and adaptations", *Ecological Economics*, Vol. 68, pp. 1915-1924.

Kenyon, W., G. Hill and P. Shannon (2008), "Scoping the role of agriculture in sustainable flood management", *Land use policy*, Vol. 25, pp. 351-360.

Kingwell, R. (2006), "Climate change in Australia: agricultural impacts and adaptation", *Australasian Agribusiness Review*, Vol. 14, No. 1.

Klein Tank, A.M.G. and G.P. Können (2003), "Trends in Indices of Daily Temperature and Precipitation Extremes in Europe, 1946-99", *Journal of Climate*, Vol. 16, pp. 3665-3680.

Knutson, T.R. and R.E. Tuleya (2004), "Impact of CO_2-induced warming on simulated hurricane intensity and precipitation: sensitivity to the choice of climate model and convective parametrization", *Journal of Climate*, Vol. 17, No. 18.

Komatsu, E., Fukushima, T. and H. Harasawa (2007), "A modeling approach to forecast the effect of long-term climate change on lake water quality", *Ecological Modelling*, Vol. 209, No. 2–4, pp. 351–366.

Kundzewicz, Z.W., L.J. Mata, N. Arnell, P. Döll, P. Kabat, B. Jiménez, K. Miller, T. Oki, Z. Şen and I. Shiklomanov (2007), Freshwater resources and their management. Climate Change 2007: Impacts, Adaptation and Vulnerability.Contribution of Working Group II to the Fourth Assessment Report of the Intergovernmental Panel on Climate Change (ed. by M.L. Parry, O.F. Canziani, J.P. Palutikof, P.J. van der Linden and C.E. Hanson), 173–210. Cambridge University Press, UK. www.ipcc.ch/pdf/assessment-report/ar4/wg2/ar4-wg2-chapter3.pdf.

Lehner, B., Döll, P., Alcamo, J. Henrichs, T. and F. Kaspar (2006), "Estimating the impact of global change on flood and drought risks in Europe: a continental, integrated analysis", *Climatic Change*, Vol. 75, pp. 273-299.

Lin, E., X. Wei, J. Hui, X. Yinlong, L. Yue, B. Liping and X. Liyong (2005), "Climate change impacts on crop yield and quality with CO_2 fertilization in China", *Philosophical Transactions of The Royal Society. Biological Science*, Vol. 360, No. 1463, pp. 2149-2154.

Lobell, D.B. and M.B. Burke (2008), "Why are agricultural impacts of climate change so uncertain? The importance of temperature relative to precipitation", *Environmental Research Letters,* Vol. 3(3).

Lobell, D.B. and C.B. Field (2007), "Global scale climate–crop yield relationships and the impacts of recent warming", *Environmental Research Letters* **2**(1).

Lobell, D., W. Schlenker and J. Costa-Roberts (2011a), "Climate trends and global crop production since 1980", *Science,* Vol. 333(6042).

Lobell, D.B., M. Bänziger, C. Magorokosho and B. Vivek (2011b), "Nonlinear heat effects on African maize as evidenced by historical yield trials", *Nature Climate Change*, Vol. 1(1), 42–45.

Lobell, D.B., A. Torney and C.B. Field (2011c), "Climate extremes in California agriculture", *Climatic Change*, Vol. 109, No. 1 Suppl., pp. 355-363.

Ludwig, F. and S. Asseng (2006) "Climate change impacts on wheat production in a Mediterranean environment in Western Australia", *Agricultural Systems*, Vol. 90, pp. 159-179.

Luo, Q., W. Bellotti, M. Williams and B. Bryan, (2005a), "Potential impact of climate change on wheat yield in South Australia", *Agricultural and Forest Meteorology*, Vol. 32, pp. 273-285.

Luo, Q., B. Bryan, W. Bellotti and M. Williams (2005b), "Spatial analysis of environmental change impacts on wheat production in mid-lower north, South Australia", *Climatic Change*, Vol. 72, pp. 213-228.

Marino, P.G., D.P. Kaiser, L. Gu and D. M. Ricciuto (2011), "Reconstruction of false spring occurrences over the southeastern United States, 1901–2007: an increasing risk of spring freeze damage?", *Environmental Research Letters*, Vol. 6, pp. 024015.

McIsaac, G.F., M.B. David, and C.A. Mitchell (2010), "Miscanthus and switchgrass production in Central Illinois: Impacts on hydrology and inorganic nitrogen leaching", *Journal of Environment* Qual. 39, pp. 1790–1799.

Meehl, G.A. and C. Tebaldi (2004), "More Intense, More Frequent, and Longer Lasting Heat Waves in the 21st Century", *Science*, Vol. 305, No. 5686, pp. 994-997.

Meehl, G.A., F. Zwiers, J. Evans, T. Knutson, L. Mearns and P. Whetton (2000), "Trends in extreme weather and climate events: issues related to modeling extremes in projections of future climate change", *Bulletin of American Meteorological Society*, Vol. 81, No. 3.

Mendelsohn, R., W.D. Nordhaus and D. Shaw (1994), "The impact of global warming on agriculture: A Ricardian analysis", *The American Economic Review*, Vol. 84, No. 4, pp. 753–771.

Messner, F., E. Penning-Rowsell, C. Green, V. Meyer, S. Tunstall, A. van der Veen, (2007) "Evaluation of flood damages: guidance and recommendations on principles and methods," FLOODsite report T09-06-01. Available at: http://www.floodsite.net

Minguez, M.I., M. Ruiz-Ramos, C.H. Díaz-Ambrona, M. Quemada and F. Sau (2007), "First-order impacts on winter and summer crops assessed with various high-resolution climate models in the Iberian Peninsula," *Climatic Change*, Vol. 81, No. 1, pp. 343-355.

Mishra, A. K. and V. P. Singh (2010), "A review of drought concepts*", Journal of Hydrology*, Vol. 391, No. 1-2, pp. 202-216."

Mitra, J. (2001), "Genetics and genetic improvement of drought resistance in crop plants". *Current Science*, Vol. 80, pp.758-762.

Moss, B., D. Stephen, D, M. Balayla, E. Becares, S.E. Collings, C. Fernandez-Alaez, M. Fernandez-Alaez, C. Ferriol, P. Garcia, J. Goma, M. Gyllstrom, L.A. Hannson, J. Hietala, T. Kairesalo, R. Miracle, S. Romo, J. Rueda, V. Russell, A. Stahl-Delbanco, M. Svensson, K. Vakkilainen, M. Valentini, W.J. van den Bund, E. van Donk, E. Vicente and M.J. Villen (2004), "Continental scale patterns of nutrient and fish effects on shallow lakes: synthesis of a pan-European mesocosm experiment", *Freshwater Biology*, Vol. 49, No. 13, pp. 1633–1649.

Mpelasoka, F., K. Hennessy, R. Jones and B. Bates (2008), "Comparison of suitable drought indices for climate change impacts assessment over Australia towards resource management*", International Journal of Climatology*, Vol. 28, pp. 1283–1292.

Murdoch, P.S., J.S. Baron, and T.L. Miller (2000), "Potential effects of climate change on surface-water quality in North America", *Journal of the American Water Resources Association*, Vol. 36(2), pp. 344-366.

Nardone, A., B. Ronchi, N. Lacetera, M.S. Ranieri and U. Bernabucci (2010), "Effects of climate changes on animal production and sustainability of livestock systems", *Livestock Science*, Vol. 130, pp. 57-69.

National Research Council (NRC) (2007), *Water Implications of Biofuels Production in the United States*, Committee on Water Implications of Biofuels Production in the United States, Water Science and Technology Board, Division on Earth and Life Studies, NRC, Washington DC.

Nelson, G.C., M.W. Rosegrant, J. Koo, R. Robertson, T. Sulser, T. Zhu, C. Ringler, S. Msangi, A. Palazzo, M. Batka, M. Magalhaes, R. Valmonte-Santos, M. Ewing and D. Lee (2009), *Climate Change, Impact on Agriculture and Costs of Adaptation*, IFPRI, Washington D. C.

Ng, T.L., J.W. Eheart, X. Cai and F. Miguez (2010), "Modeling miscanthus in the Soil and Water Assessment Tool (SWAT) to simulate its water quality effects as a bioenergy crop". *Environmental Science and Technology*, 44 (18), pp. 7138–7144.

OECD (2013a), *Water and Climate Change Adaptation: Policies to Navigate Uncharted Waters*, OECD Studies on Water, OECD Publishing, Paris.
doi: 10.1787/9789264200449-en.

OECD (2013b), *OECD Compendium of Agri-environmental Indicators*, OECD Publishing, Paris.
doi: 10.1787/9789264186217-en

OECD (2012a), *OECD Environmental Outlook to 2050: The Consequences of Inaction*, OECD Publishing, Paris.
doi: 10.1787/9789264122246-en.

OECD (2012b),*Water Quality and Agriculture: Meeting the Policy Challenge*, OECD Studies on Water, OECD Publishing, Paris.
doi: 10.1787/9789264168060-en.

OECD (2010), *Climate Change and Agriculture: Impacts, Adaptation and Mitigation*, OECD Publishing, Paris.
doi: 10.1787/9789264086876-en.

Ortiz-Bobea (2012), "Understanding Heat and Moisture Interactions in the Economics of Climate Change Impacts and Adaptation on Agriculture", Working paper.

Ortiz-Bobea, A. and R.E. Just (2012), "Modeling the structure of adaptation in climate change impact assessment", *American Journal of Agricultural Economics*, published online.

Pantaleoni, E., B.A. Engel and C.J. Johannsen (2007), "Identifying agricultural flood damage using Landsat imagery", *Precision Agriculture*, Vol. 8, No. 1-2, pp. 27–36.

Park, T., C. Ho, S. Jeong, Y. Choi, S.K. Park and C. Song (2011), "Different characteristics of cold day and cold surge frequency over East Asia in a global warming situation", *Journal of Geophysical Research: Atmospheres*, Vol. 116, No. D12.

Parry, M.L., C. Rosenzweig, A. Iglesias, M. Livermore and G. Fischer (2004), "Effects of climate change on global food production under SRES emissions and socio-economic scenarios", *Global Environmental Change*, Vol. 14, pp. 53-67.

Parsons, D.J., A.C. Armstrong, J.R. Turnpenny, A.M. Matthews, K. Cooper and J.A. Clark (2001), "Integrated models of livestock systems for climate change studies. 1. Grazing systems", *Global Change Biology*, Vol. 7, pp. 93-112.

Philpott, S.M., B.B. Lin, S. Jha and S.J. Brines (2008), "A multi-scale assessment of hurricane impacts on agricultural landscapes based on land use and topographic features", *Agriculture, Ecosystems and Environment*, Vol. 128, No. 1-2, pp. 12-20.

Piao, S.L., J.Y. Fang, L.M. Zhou, P. Ciais and B. Zhu (2006), "Variations in satellite-derived phenology in China's temperate vegetation", *Global Change Biology*, Vol. 12, pp. 672–685.

Piao, S., P. Ciais, Y. Huang, Z. Shen, S. Peng, J. Li, H. Liu, Y. Ma, Y. Ding, P. Friedlingstein, C. Liu, K. Tan, Y. Yu, T. Zhang and J. Fang (2010), "The impacts of climate change on water resources and agriculture in China", *Nature*, 467, pp. 43-51.

Pivot, J., E. Josien and P. Martin (2002), "Farms adaptation to changes in flood risk: A management approach", *Journal of Hydrology*, Vol. 267, No. 1-2, pp. 12-25.

Planton, S., M. Dèque, F. Chauvin and L. Terray (2008), "Expected impacts of climate change on extreme climate events", *Comptes Rendus Geoscience*, Vol. 340, No. 9-10, pp. 564-574.

Quiggin, J. and J. Horowitz (2003), "Costs of adjustment to climate change", *Australian Journal of Agricultural and Resource Economics*, Vol. 47, pp. 429-446.

Reyenga, P.J., S.M. Howden, H. Meinke and G.M. McKeon (1999), Modelling global change impacts on wheat cropping in south-east Queensland, Australia. Environmental Modelling and Software 14, 297-306.

Rigby, J.R. and A. Porporato (2008), "Spring frost risk in a changing climate", *Geophysical Research Letters*, Vol. 35, No. 12.

Roberts, M.J., W. Schlenker and J. Eyer (2012), "Agronomic weather measures in econometric models of crop yield with implications for climate change", *American Journal of Agricultural Economics*, Vol. 95, No. 2, pp. 236-243.

Rosenzweig, C. and M.L. Parry (1994), "Potential impact of climate change on world food supply", *Nature*, Vol. 367, pp. 133-138.

Rosenzweig, C., A. Iglesias, X.B. Yang, P.R. Epstein and E. Chivian (2001), "Climate change and extreme weather events. Implications for food production, plant diseases, and pests", *Global change and human health*, Vol. 2, No. 2.

Rosenzweig, C., F.N. Tubiello, R. Goldberg, E. Mills and J. Bloomfield (2002), "Increased crop damage in the US from excess precipitation under climate change", *Global Environmental Change*, Vol. 12, No. 3, pp. 197-202.

Rosenzweig C., J.W. Jones, J.L. Hatfield, A.C. Ruane, K.J. Boote, P. Thorburn, J.M. Antle, G.C. Nelson, C. Porter, S. Janssen, S. Asseng, B. Basso, F. Ewert, D. Wallach, G. Baigorria and J. M. Winter (2013), "The Agricultural Model Intercomparison and Improvement Project (AgMIP): Protocols and pilot studies", *Agricultural and Forest Meteorology*, Vol. 170, pp. 166-182.

Rötter, R. and S.C. van de Geijn (1999), "Climate change effects on plant growth, crop yield and livestock", *Climatic Change*, Vol. 43, No. 4, pp. 651-681.

Sadras, V.O. and J. P. Monzon (2006), "Modelled wheat phenology captures rising temperature trends: shortened time to flowering and maturity in Australia and Argentina", *Field Crops Research*, Vol. 99, pp. 136-146.

Schaap, B.F., M. Blom-Zandstra, C.M.L. Hermans, B.G. Meerburg and J. Verhagen (2011), "Impact changes of climatic extremes on arable farming in the north of the Netherlands", *Regional Environmental Change*, Vol. 11, No. 3, pp. 731-741.

Schlenker, W. and M. Roberts (2009), "Nonlinear temperature effects indicate severe damages to us crop yields under climate change". *Proceedings of the National Academy of Sciences*, 106(37), 15594.

Seo, N.S., B.A. McCarl and R. Mendelsohn (2010), "From beef cattle to sheep under global warming? An analysis of adaptation by livestock species choice in South America", *Ecological Economics*, Vol. 69, No. 12, pp. 2486-2494.

Seo, S.N. and B. McCarl (2011), "Managing livestock species under climate change in Australia, Animals, Vol. 1, No. 4, pp. 343-365.

Seo, S.N. and R. Mendelsohn (2008), "Measuring impacts and adaptations to climate change a structural Ricardian model of African livestock management", *Agricultural Economics*, Vol. 38, No. 2, pp. 151-165.

Shabbar, A. and B. Bonsal (2003), "An assessment of changes in winter cold and warm spells over Canada", *Natural Hazards*, Vol. 29, No. 2, pp. 173–188.

Sheffield, J., E.F. Wood and M.L. Roderick (2012), "Little change in global drought over the past 60 years", *Nature*, Vol. 491, pp. 435-438.

Sherif M.M. and V.P. Singh (1999), "Effect of climate change on sea water intrusion in coastal aquifers", *Hydrological Processes*, Vol. 13, pp. 1277-1287.

Sinclair, T. (2010), *Precipitation: The thousand-pound gorilla in crop response to climate change*, World Scientific Books, Hackensack, NJ., pp. 179–190.

Singh, S.K., H.R. Meena, D.V. Kolekar and Y.P. Singh (2012), "Climate change impacts on livestock and adaptation strategies to sustain livestock production", *Journal of Veterinary Advances*, Vol. 2, No. 7, pp. 407-412.

Smith, D.I. (1994), "Flood damage estimation – A review of urban stage damage curves and loss functions", *Water SA*, Vol. 20, No. 3.

Sonnenborg, T.O., K. Hinsby, L. van Roosmalen and S. Stisen (2012), "Assessment of climate change impacts on the quantity and quality of a coastal catchment using a coupled groundwater-surface water model", *Climatic Change*, Vol. 113, No. 3-4, pp. 1025-1048.

Tao, F.L., M. Yokozawa, J.Y. Liu and Z. Zhang (2008), "Climate-crop yield relationships at provincial scales in China and the impacts of recent climate trends", *Climate Research*, Vol. 38, pp. 83–94.

Tao, F., M. Yokozawa, Y. Hayashi and E. Lin (2003), "Future climate change, the agricultural water cycle, and agricultural production in China", *Agriculture, Ecosystems and Environment*, Vol. 95, No. 1, pp. 203-215.

Tapia-Silva, F., S. Itzerott, S. Foerster, B. Kuhlmann and H. Kreibich (2011), "Estimation of flood losses to agricultural crops using remote sensing", *Physics and Chemistry of the Earth*, Vol. 36, No. 7-8, pp. 253-266.

Taylor, R.G., B. Scanlon, P. Döll, M. Rodell, R. van Beek, Y. Wada, L. Longuevergne, M. Leblanc, J.S. Famiglietti, M. Edmunds, L. Konikow, T.R. Green, J. Chen, M. Taniguchi, M.F.P. Bierkens, A. MacDonald, Y. Fan, R.M. Maxwell, Y. Yechieli, J.J. Gurdak, D.M. Allen, M. Shamsudduha, K. Hiscock, P.J.-F. Yeh, I. Holman and H. Treidel (2012), "Ground water and climate change", *Nature Climate Change*, Vol. 3, No. 1.

Thornton, P.K., J. van de Streeg, A. Notenbaert and M. Herrero (2009), "The impacts of climate change on livestock and livestock systems in developing countries. A review of what we know and what we need to know", *Agricultural Systems*, Vol. 101, No. 3, pp. 113-127.

Tubiello, F.N., J-F Soussana and S. M. Howden (2007), "Crop and pasture response to climate change", *Proceedings of the National Academy of Sciences of USA*, Vol. 104, No. 50, pp. 19686-19690.

van der Velde, M., G. Wriedt and F. Bouraoui (2010), "Estimating irrigation use and effects on maize yield during the 2003 heatwave in France", *Agriculture, Ecosystems and Environment*, Vol. 135, No. 1-2, pp. 90–97.

van Ittersum, M.K., Howden, S.M. and Asseng, S. (2003) Sensitivity of productivity and deep drainage of wheat cropping systems in a Mediterranean environment to changes in CO_2, temperature and precipitation. Agriculture, Ecosystems and Environment, Vol. 97, 255-273.

Varanou, E., E. Gkouvatsou, E. Baltas and M. Mimikou (2002), "Quantity and quality integrated catchment modeling under climate change with use of soil and water assessment tool model", *Journal of Hydrologic Engineering*, Vol. 7, No. 3, pp. 228–244.

Wang, D., M. Hejazi, X. Cai, and A. J. Valocchi (2011), "Climate change impact on meteorological, agricultural, and hydrological drought in central Illinois", *Water Resources Research*, Vol. 47, No. 9, p. W09527.

Wang, G. (2005), "Agricultural drought in a future climate: results from 15 global climate models participating in the IPCC 4th assessment", *Climate Dynamics*, Vol. 25, No. 7, pp. 739-753.

Wang, J., R. Mendelsohn, A. Dinar, J. Huang, S. Rozelle and L. Zhang (2009a), "The impact of climate change on China's agriculture", *Agricultural Economics*, Vol. 40, No. 3, pp. 323–337.

Wang, J., E. Wang, Q. Luo and M. Kirby (2009b), "Modelling the sensitivity of wheat growth and water balance to climate change in southeast Australia", *Climatic Change*, Vol. 96, pp. 79-96.

Werner, A.D., M. Bakker, V.E.A. Post, A. Vandenbohede, C. Lu, B. Ataie-Ashtiani, C.T. Simmons and D.A. Barry (2012), *Seawater intrusion processes, investigation and management: Recent advances and future challenges*, Advances in water resources, In Press, http://dx.doi.org/10.1016/j.advwatres.2012.03.004.

Whitehead, P.G., R.L. Wilby, D. Butterfield and A.J. Wade (2006), *Impacts of climate change on in-stream nitrogen in lowland chalk streams: an appraisal of adaptation strategies.* Science Total Environment, Vol. 365, No. 1-3, pp. 260–273.

Whitehead, P.G., R.L. Wilby, R.W. Battarbee, M. Kernan and A.J. Wade (2009), "A review of the potential impacts of climate change on surface water quality", *Hydrological Sciences Journal*, Vol. 54, No. 1, pp. 101-123.

Wilby, R.L. (1994), "Exceptional weather in the Midlands, UK during 1988–1990 results in the rapid acidification of an upland stream", *Environmental Pollution*, Vol. 86, pp. 15–19.

Yao, F., P. Qin, J. Zhang, E. Lin and V. Boken (2011), "Uncertainties in assessing the effect of climate change on agriculture using model simulation and uncertainty processing methods", *Chinese Science Bulletin*, Vol. 56, No. 8, pp. 729-737.

Zhang, X. and X. Cai (2013), "Climate Change Impacts on Global Agricultural Water Deficit", *Geophysical Research Letters*, Vol. 40, Issue 6, pp. 1111-1117.

Zhou, G., S. Wan, G. Feng and W. He (2012), "Effects of regional warming on extreme monthly low temperatures distribution in China", *International Journal of Climatology*, Vol. 32, pp. 387–391.

Chapter 2

Climate change adaptation and mitigation strategies for agricultural water management

The present chapter proposes an economic framework for analysing the adaptation of agricultural water management to climate change, and examines the role and capacity of different policy instruments to foster the adaptation of agricultural water management to climate change.

The statistical data for Israel are supplied by and under the responsibility of the relevant Israeli authorities. The use of such data by the OECD is without prejudice to the status of the Golan Heights, East Jerusalem and Israeli settlements in the West Bank under the terms of international law.

The previous chapter has shown that climate change is projected to have significant impacts on water systems and agricultural production. However, the major challenge for policy makers is that the accuracy of existing evidence on projected climate impacts on water is quite low. Climate projections are typically very uncertain, especially for precipitation, and downscaling is fraught with problems. In other words, projections at the resolution and scale required for adaptation decisions are lacking. Adaptation decisions thus need to accommodate considerable uncertainty. The present chapter builds on this fundamental characteristic to propose an economic framework for analysing adaptation of agricultural water management to climate change, and discuss the role and capacity of different policy instruments to foster the adaptation of agricultural water management to climate change.

Adaptation policies for agricultural water management: An economic framework

The features of the adaptation problem: uncertainty, long-term and complex interactions

Adaptation to climate change can be defined as "an adjustment in ecological, social or economic systems in response to observed or expected changes in climatic stimuli and their effects and impacts, in order to alleviate adverse impacts of change or take advantage of new opportunities" (OECD, 2010b). Other definitions can be found in the literature and policy analysis (see in particular Hallegatte et al., 2011), but all underline the central issue of adaptation, which is to assess the effectiveness of adaptation options at attenuating the impacts of climate change and their associated costs. Two dimensions are especially important for analysing adaptation strategies: the first one is the presence of deep uncertainty and timing issues, which complicate the task of decision makers; the second one is the need to clarify the specific role of government intervention.

The presence of *deep uncertainty*[1] makes it difficult to determine the optimal sequence of actions and their timing. Previous chapters have shown there are many impacts of climate change on the water cycle as well as on the associated consequences for agriculture which are very difficult to predict, in particular at the local scale where most adaptation decisions take place. Moreover, the way the different climatic variables – temperature, precipitation, evapotranspiration – interact to affect jointly agricultural production is not fully understood, rendering the choice of adaptation responses difficult. Investing today in adaptation is costly, and the expected benefits of such investment are far off in time – as their weight is reduced by the discount rate – and very uncertain. Under these circumstances, the incentive to invest in adaptation is naturally low for both agents and institutions. Government could apply a lower social discount rate when assessing present discounted values of future benefits and costs. However, determining the appropriate social discount rate is itself highly uncertain and raises ethical problems. Indeed, it is not just a technical issue of estimating short-run time preferences, but also a matter of inter-temporal equity between generations (Gollier, 2013; Fleurbay and Zuber, 2012).

In such a delicate context, rational economic actors and institutions are naturally incited to adopt a dynamic learning approach that anticipates in a continuous timeframe the arrival of new information on climate change impact assessment, more detailed assessment of regional or local impacts, and the development of new technologies able to deal more efficiently with climate change conditions, etc. Adaptation to climate change is a continuous process, involving learning and revisions of beliefs, rather than a discrete choice between a well-defined set of options. It combines reactive adaptation and anticipatory actions based on the present state of knowledge and expectations of decision makers, in a way that can make the distinction most often meaningless (Smit et al., 2000). In these circumstances, investments that increase the ability of economic agents and institutions to react more quickly and smoothly to on-going changes are of crucial importance. A central question is whether the "signal" of climate change is strong enough to have enough impact on these decisions.

Under deep uncertainty, adaptation strategies thus require a constant balance between the need to tackle a problem and the existing uncertainties (**Box 2.1**). However, as underlined in Chapter 1, when assessing the costs and benefits of adaptation options, one should also take into account other major drivers such as the increase in food demand in the next century. Water is a critical resource essential to life and ecosystems, and socio-economic developments over the coming decades justify on their own improvement of water management and water use efficiency. In other words, even without climate change, one can expect that the relative (shadow) price of water will increase. Taking into account this trend towards an increase in the shadow prices of natural resources such as water due to increasing expected scarcity, can counterbalance the role of the discount rate in cost-benefit analysis for long-term horizons (OECD, 2006). This is the sense of "no-regrets" adaptation options.

Box 2.1. Economic approaches for evaluating adaptation measures: Advantages and limitations

Performance criteria for judging adaptation measures are needed to guide *ex ante* policy measure choice and design decisions and to measure *ex post* policy performance. Adaptation objectives should be set and achieved with economic efficiency in mind, such that: (i) the marginal benefits and costs of achieving the adaptation objectives should be reasonably balanced; and (ii) whatever adaptation objective is set, the objective should be achieved at least cost. Three criteria for good adaptation policy are *effectiveness*, *economic efficiency* and *equity* (Cimato and Mullan, 2010). Effectiveness refers to the capacity of the instruments to achieve stated adaptation goals whereas economic efficiency is promoted by selecting policy instrument that minimises compliance costs while achieving adaptation goal, thus maximising cost-effectiveness. A final criterion that plays an important role in the choice and evaluation of adaptation policy is the equity of the distribution of economic costs and benefits between and among different groups.

With regard to policy evaluation, social cost-benefit analysis is closest to a social welfare analysis (Johansson, 1991). However, social cost-benefit analysis is a very information-intensive methodology raising considerable methodological and measurement challenges, since monetary estimates for non-market goods are needed. The basic idea behind cost-benefit analysis is to measure in monetary units how social welfare is affected by a particular adaptation policy. Cost-benefit analysis can be done either *ex ante* or *ex post*. The rationale for *ex ante* analysis is that it will provide information as to whether the proposed adaptation policy is socially profitable or not. *Ex post* analysis assists the process of learning about what does and does not contribute to overall social well-being (OECD, 2006).

Cost-benefit analysis of adaptation policy is difficult because of the uncertainty related to the impacts of climate change and thus benefits of adaptation measures. Hallegatte et al. (2012) argue that despite the uncertainty, the cost-benefit analysis remains a reference method. Moreover, one can use benefit-cost analysis with at least two "optimistic" and "pessimistic" scenarios with occurrence probabilities, while being careful to check the robustness of results as regards the choice of probabilities (Hallegatte et al., 2012).

Because climate impacts are highly uncertain, the adaptation policy benefits —and sometimes costs — are uncertain with unknown outcomes and probabilities. Thus, one may need to use decision-making approaches that help to select policy measures that are robust to these uncertainties (Cimato and Mullan, 2010). A primary example of these approaches is the so-called minimax, in which one minimises the possible loss related to maximum loss scenario.

Since climate change adaptation is a dynamic process, adaptation policies need to allow for flexibility and learning in so far as possible (Cimato and Mullan, 2010). One possible solution is to employ the Real Options Approach (ROA) that allows the inclusion of flexibility (the value of waiting) as part of the costs of the investment (Pindyck, 1989; Dixit and Pindyck, 1994). The ROA provides a dynamic learning mechanism that allows the opportunity to phase in investments and stage key decisions. However, in the context of adaptation investments, the ROA has some caveats, e.g. as regards the stochastic nature of climate patterns (uncertainty is not necessarily reduced over time) and potential irreversible damage owing to delaying decisions (Cimato and Mullan, 2010).

The purposes of adaptation strategies: Reducing vulnerability, increasing resilience

In this context of deep uncertainty, it is generally recognised that the response in terms of adaptation strategies should focus more on the *adaptive capacities* of systems than adaptation choices themselves. Climate change being a continuous process, potentially subject to surprises, limiting the problem of adaptation to the question of a choice between predefined technical introduces excessive rigidity in problem solving. The objective should better focus on reducing the overall vulnerability of the system impacted, i.e. the "propensity or predisposition to be adversely affected" (IPCC, 2012). Vulnerability comes from a combination of a given level of exposure and of sensitivity to climate change and associated shocks (**Figure 2.1**). Hence reducing exposure and sensitivity, as well as increasing adaptive capacities of the system are the three main ways of addressing vulnerability to climate change. A closely related objective of adaptation strategies is to improve *resilience*, which can be defined as the "capacity of systems, communities, households or individuals to prevent, mitigate or cope with risk and recover from shocks" (Gitz and Meybeck, 2012). Resilience is strongly linked to the reduction of vulnerability, but, as underlined by Gitz and Meybeck, it also includes the idea of recovery from shocks, and so has a more dynamic dimension than vulnerability, which is more static. The resilience of systems thus deals with the ability to continuously adapt to changing climate conditions, to absorb related shocks, and to recover a path towards growth and development.

Vulnerability and resilience do not just concern the physical impacts of climate change on agricultural systems, but also address economic, social and environmental impacts of climate change. Making the system less vulnerable and more resilient requires taking into account the different levels of actions of the system, and their interactions. For example, suppose that in a given region climate change increases the frequency and severity of droughts, and that these changes could not be fully anticipated by farmers. Hence production systems, if not adapted, may suffer from more frequent and substantial crop yield losses. If crop insurance is available, it can improve the resilience of farmers by mitigating the huge income losses due to these more frequent weather shocks, and thus recover from them more easily. It can also provide him a price signal through the insurance premium indicating the rising cost of risk, providing incentives to reduce risk exposure in the longer run by adapting cropping systems.

But reducing vulnerability to changing climate does not always lead to increasing resilience, nor is ensuring the sustainability of the system in the long run. Diminishing vulnerability in the short-run can even increase vulnerability in the long-run. For example, in order to mitigate water supply risk, irrigators could be incited to move from surface water, whose supply becomes more and more volatile due to climate change, to groundwater reserves for their freshwater withdrawals. But in the cases of non-renewable groundwater resources, this strategy is more like a "headlong rush" rather than a sustainable adaptation to climate change. More importantly, such temporary access to groundwater can also undermine farmers' incentives to invest in water use efficiency, and finally delay adaptation. A similar line of reasoning can apply to government subsidised insurance and compensation systems that are not based on fair insurance premiums. They allow reducing vulnerability in the short-run and recovering from shocks, but without an appropriate price signal for risk taking, they can increase risk exposure and delay adaptation in the longer run. In a similar vein, some authors underline the importance of the distinction between resilience and *resistance* (Dauphiné and Provitolo, 2007). Increasing resistance mainly consists in reducing vulnerability to shocks, while resilience puts more emphasis on improving the capacity of a given system to recover its initial state, and involves qualities such as diversification, self-organisation and learning.

Figure 2.1. Vulnerability and resilience

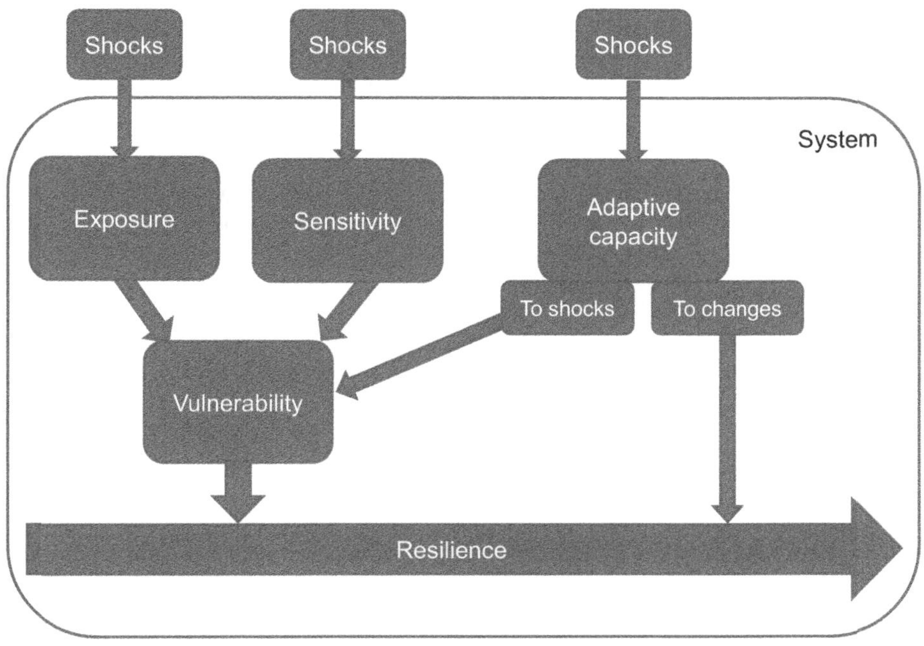

Source: Gitz, V. and A. Meybeck (2012), "Risks, Vulnerabilities and Resilience in a Context of Climate Change", in Meybeck, A., Lankoski, J., Redfern, S., Azzu, N. and V. Gitz (eds.) *Building resilience for adaptation to climate change in the agriculture sector – Proceedings of a Joint FAO/OECD Workshop*, Rome, 23-24 April 2012.

Economic rationale for government policy intervention in climate change adaptation

In most OECD countries, governments or local authorities already play an important role in agricultural water management, through innovation policies, water management planning, water policy instruments, compensation and insurance mechanisms against natural disasters, and more broadly agricultural and agri-environmental policies. It therefore seems natural that government policies are involved in adaptation to climate change. However, the way adaptation modifies the classical role of government policies in these domains deserves discussion.

It is in fact important to clarify the reasons that can justify specific government intervention in adaptation strategies to climate change. Hallegatte et al. (2011) propose an informative economic analysis of the rationales for climate change adaptation policies. According to these authors, there is a fundamental economic difference between mitigation and adaptation: while mitigation is attached to the preservation of a public good, which traditionally implies some form of policy intervention, adaptation tends in practice to be associated with private goods and/or services. Examples of such private goods in the agricultural sector include: investment in a more water efficient irrigation technology; changes in agricultural management practices such as crop mixes and sowing dates; and soil testing.

From a public policy perspective, considering adaptation as a matter of private goods and services tends to support *laissez-faire* as the more efficient way to meet the adaptation challenge. In this framework, each farmer would individually invest in adaptation practices according to the current state of knowledge available, expectations of future climates, etc., in order to maximise his own utility, and the collective outcome of these decentralised climate change adaptation choices would coincide with the social welfare optimum.

According to Hallegatte et al. (2011), this "private goods approach" of the adaptation issue does not take into account the specific features that characterise adaptation, which are summarised in **Box 2.2**. Some of these features are especially important in the context of agricultural water management – and water management in general – and strongly support dedicated climate change adaptation policies, in particular:

- *Water as a common pool resource:* although situations vary a great deal among countries, it is widely recognised that water has often specific features that make it difficult to define property rights, in particular the lack of excludability. When water resources are indeed common pool, there is a need for coordination mechanisms between users in order to avoid its economically inefficient overconsumption.

- *The* existence *of external effects on water systems*: Agriculture affects water quality through nutrient runoff and leaching – nitrogen and phosphorous – as well as pesticides (OECD, 2012b).

- *The presence of infrastructure networks*: This is the case of irrigation systems and, more generally, water networks.

Box 2.2. Features of adaptation that may require policy intervention

The following characteristics of adaptation are susceptible to discentivise producers and/or consumers to invest adequately in adaptation measures, i.e. both from private and social points of view.

1. Poor dissemination of available information

2. Barriers to collective action at the local level.

3. Decision routines and inadequate consideration of long-term consequences on private investment decisions.

4. Negative or positive external impacts.

5. The role of major infrastructure networks for the public benefit.

6. Inadequacy of existing standards and regulations.

7. Poverty and budget constraints.

Source: Hallegatte, S., A. Shah, R. Lempert, C. Brown, S. Gill (2012), ""Investment Decision Making Under Deep Uncertainty – Application to Climate Change", *Policy Research Working Paper No. 6193*, World Bank, Washington DC.

These features call for specific public policy interventions that are not unique to the issue of adaptation. External effects and common property resources are already an essential dimension of agricultural water management, even without considering adaptation. What is at stake, however, is the way adaptation choices can affect the levels of these external effects and common property resources. In sum, the real issue is the *interaction between already existing market failures and adaptation choices*. Some adaptation choices can worsen or improve external effects, as they can worsen or improve the overconsumption of water resources. For example, irrigation can be an adaptation response to water deficit at the farm level, but if all farmers of a given watershed move from rainfed to irrigated agriculture, this would worsen the overconsumption of water.

Accounting for the market failures related to water management described above, it is thus clear that *private vulnerability* to climate change is not synonymous with *social vulnerability*. Similarly, *private resilience* is not equivalent to *social resilience*. Thus the role of government policies in the domain of agricultural water management could be to ensure that resilient adaptation strategies are aligned as far as possible with the overall objective of social welfare, including the environmental and social dimensions.

The different levels of action for the adaptation of agricultural water management to climate change

In order to structure the policy discussion on the basis of the framework presented above, the following sections will consider the five following levels of action for analysing adaptation strategies of agricultural water management.

- On-farm adaptation of water management.

- Water management policies for adaptation at the watershed level.

- Risk management approaches for adapting to increasing risks of droughts and floods.

- Agricultural policy coherence and the role of market drivers.

- Interactions between mitigation and adaptation of agricultural water management.

The remainder of this chapter will be devoted exclusively to the analysis of each of these five levels of action. The main policy recommendations arising from the following analysis will be presented in Chapter 3.

On-farm adaptation of water management

The first level of adaptation is the farm. On-farm adaptation actions are usually defined as changes in agricultural management practices – eventually the whole cropping and livestock systems – leading to a reduction of the impacts or are related to new patterns of weather variables. Of course farmers' decisions to adapt depend on many drivers such as policies and markets, environmental regulations, and institutions. Some of these drivers may facilitate or, on the contrary delay the adoption of on-farm adaptation practices. This is an important issue, and is discussed later in this chapter. The purpose of this section is to focus on the main adaptations of agricultural management practices to climate change, and to understand how public policies can foster their adoption by farmers, in a given policy and market environment.

Reviewing the main adaptation options for agricultural water management

As regards crops, the main adaptation options at the farm level to climate change include the following (see FAO, 2011 for an extensive review of technical options; Saleth et al., 2011 in the case of drought).

- Adoption of drought-resistant varieties.

- Change in sowing dates to benefit from a longer growing season and reduce the probability of the crop being exposed to a drought period.

- Increased irrigation efficiency to reduce the sensitivity of the farm to swing in water supply conditions.

- Adoption of irrigation in previously non-irrigated agricultural areas to overcome a water deficit.

- Changes in crop rotations to include crops that are less exposed and/or less sensitive to water deficits or droughts, etc.

- Changes in agricultural practices, such as adoption of conservation tillage and agroforestry.

For livestock, adaptation measures can relate to feed production, livestock water requirements and animal health (heat stress and diseases). Below, the main adaptation options are presented successively with a focus on crop production. Crop and livestock adaptation options are closely interrelated, as both livestock and feed production are dependent on water resources and quality.

Drought-resistant varieties

A major issue for the adaptation of crop varieties to climate change is resistance to drought. Although it seems clear that plant varieties will be bred to change with the climate, the degree to which these adaptations will mitigate damages is uncertain and highly dependent on both the degree of warming and uncertain changes in precipitation patterns.

Tolerance to heat and drought generally comes at the expense of yield potential. One mechanism bred into certain varieties of maize, causes the plant's stoma to close when vapor pressure deficit and evapotranspiration increase (Sinclair and Muchow, 2001). This mechanism has the benefit of preserving soil moisture and reducing the chance of severe water stress in times of drought. The cost is that photosynthesis shuts down when the stoma close, reducing yield potential. Such varieties can improve yield in water stressed environments, but the tradeoff is a delicate one.

Another standard mechanism for adapting plants to particular climates pertains to its rate of maturity. As the growing season become longer with climate change, it may be beneficial to breed plants that mature more slowly, allowing them to absorb and transform sunlight into growth and yield. The tradeoff with having a slower maturing plant is that it may become more susceptible to extreme heat and drought. For example, in the southern US states, growing seasons are generally longer, but maturing rates are actually selected to be shorter, so that full maturity comes about before late summer heat and dryness can damage the plant.

Relatively new, genetically modified crops may improve drought tolerance by having deeper roots that can absorb more moisture from soils. Evidence that these varieties actually improve drought tolerance has had limited empirical support, in large part because these varieties are relatively new and, until very recently, the weather has been temperate. Moreover, such modified root systems, by being able to absorb more water, could also put further pressure on already scarce water reserves.

Adjusting planting times

Planting times may be adjusted earlier to take advantage of longer growing seasons and crops bred with slower rates of maturity. Earlier planting times could bring the critical flowering period earlier to reduce exposure to mid-summer extreme heat. There is some evidence that such planting adjustments could offset damages from extreme heat (Ortiz-Bobea and Just, 2012; Berry et al., 2012; Butler and Huybers, 2013). **Table 2.1** illustrates the potential economic savings from this adaptation option in the case of the United States. These savings would range from tens to hundreds of millions of dollars, depending on the State considered.

There are, however, remaining challenges and unknowns. For example, recent experience in 2012 revealed that US corn, which had record early plantings, was still severely damaged by the summer's extreme heat that arrived during a particularly sensitive period for the plant (Berry et al., 2012). Another issue is that, if it were to be used, this approach would apply mostly to wind-pollinated crops and might not work for insect pollinated crops. Indeed, shifting planting times could eventually result in mismatch timing with pollinators. These challenges require a continuous stream of research and learning-by-doing process to assess the potential for mitigating damages of extreme heat via adjustments to maturity and planting times.

Table 2.1. Corn yield impacts from a uniform 5 F warming and influence of planting date as an adaptation response

	Without change in planting date (%)/(bu/acre)	With change in planting date (%)/(bu/acre)	Impact mitigated with adaptation (%)	Optimal change in planting date (days)	Savings from adaptation (million 2010 USD)
Illinois	-34.7/-47.3	-21.9/-29.9	36.9	-16	1 371
Indiana	-26.8/-35.3	-14.9/-19.6	44.4	-18	405
Iowa	-27.1/-37.2	-18.0/-24.7	33.6	-14	848
Michigan	-19.2/-21.6	-6.6/-7.5	65.3	-18	168
Minnesota	-20.6/-26.8	-11.2/-14.6	45.5	-14	330
Ohio	-21.4/-27.0	-10.4/-13.1	51.4	-17	116
Pennsylvania	-23.9/-24.7	-7.0/-7.2	70.6	-20	61
Wisconsin	-17.3/-20.8	-7.4/-8.9	56.8	-15	102
Full sample	-26.3/-34.4	-14.0/-18.5	44.1	-15.8	3 401

Source: Ortiz-Bobea, A. and R.E. Just (2012), "Modeling the structure of adaptation in climate change impact assessment", *American Journal of Agricultural Economics*, published online.

Irrigation

As shown in Chapter 1, climate change could increase crop water requirements in several regions of the world, which would imply an increasing demand for irrigation by farmers. Also more frequent and severe droughts could incite farmers to invest in securing their access to water resources, in view of mitigating yield losses due to insufficient rainfall. For example, the role of irrigation as a farm adaptation response in interaction with crop technology is underlined by Fleischer and Kurukulasuriya (2012) in the case of **Africa** and **Israel**.

However, the potential of irrigation as an on-farm adaptation strategy depends on the availability of water resources, which could be reduced by climate change and rising competition between water users. Many regions that are currently heavily irrigated may have severely reduced availability of irrigation under climate change. Most heavily irrigated areas, like **California**'s Central Valley, rely on surface water delivered via rivers and canals from a network of reservoirs that capture approximately 50% of all precipitation in the state. Snow pack in the Sierra Mountains also acts as a form of storage. With climate change, an earlier spring melt will effectively reduce natural storage in the form of snow pack. Similar problems are likely in other parts of the world that rely on snow and glacier melts for irrigation water (Barnett et al., 2005). In many parts of the world, surface and groundwater sources are being depleted, in part because of rapidly growing populations, changing land use, common pool resource problems, as well as subsidies for water collection, storage, and delivery.

In general, there are likely large potential gains in water use efficiency from improved irrigation systems, land management practices for preparing fields for efficient irrigation and managing excess water, and possibly reallocation of land to different uses.

Shifting crops and new cropping areas

As described previously, it seems likely that crop choices will change in relation to changing growing conditions. There has been some preliminary empirical work that attempts to estimate crop choice in response to climate and climate change (e.g. Seo and Mendelsohn, 2008). New arable cropping areas in northern latitudes, especially in northern Eurasia, may also open up to offset losses in warmer regions. This is an area that requires more research, and should be combined with partial and general equilibrium models to account for price and crop choices simultaneously. Most computational models to date hold growing cultivated areas largely fixed, or allow for anticipated losses of cropland but not for potential gains (Nelson, 2009).

Combining farm practices in a holistic agronomic approach

The on-farm adaptation options presented above can widely differ in terms of costs and benefits. In general, a change in sowing dates is considered a low cost option. Conversely, investing in irrigation on a previously rainfed cropping system usually requires substantial infrastructural investments. Assessing the costs and benefits of adaptation strategies in agriculture is however a complex task that involves understanding and characterising the relationships between agricultural management practices, weather variables and production. On-farm adaptation to climate change cannot be reduced to the adoption of a single technique in isolation, but will require farmers to reconsider in a coherent and holistic manner their production systems, and the interactions between their different components. There are far more technical options that the ones considered above, and even more combinations of agricultural management practices.

For example, a recent study from Brisson et al. (2010) has tried to disentangle the different factors explaining the stagnation of wheat yields in Europe in the recent decades, by using a mix of three approaches: national and regional statistics, scattered trials, and results of agro-climatic models using climatic data. Their results tend to show that during the recent decades, genetic progress has still increased, but this positive factor have been counteracted by a tendency towards unfavourable climate conditions for crop growing. Changes in agricultural management practices also play a role, according to this study, although these are less significant than genetic improvement and climate.

Other works by Ortiz-Orbea and Just (2012) and Ortiz-Orbea (2012) also underlines the importance of an integrated agronomic approach to evaluate the costs and benefits of adaptation options, taking into account the relative roles of temperature and soil moisture in explaining yields statistically; and focussing on crucial phenological phases of crop growth such as the flowering period. If there is water stress at this stage of plant development, even if water availability is sufficient the rest of the time, one can expect substantial crop yield losses. More integrated approaches for evaluating adaptation options at the agricultural sector level consist in combining agronomic models of agricultural production with integrated models of watersheds and economic models, to produce an integrated assessment of adaptation options.

In recent years, several OECD countries have undertaken integrated evaluation studies in order to assess the impacts of climate change on the main sectors of their economy, including agriculture. This is for instance the case of the **European Union** with the report *Climate Change, impacts and vulnerability in Europe 2012* (EEA, 2012) and the United States, with a specific report devoted to the agricultural sector (see **Box 2.3** for a summary of the main outcomes). Another example of integrated approach is the IMPACT model that studies the impacts of climate change on water for rainfed and irrigated agriculture at the world level, and allows for analysing the roles of adaptation options such as improving agricultural water use efficiency and developing irrigation (**Box 2.4**). A major advantage of such an approach is to include supply and demand for food and for water, which can be influenced by adaptation options and thus could affect their costs and benefits.

Box 2.3. Agricultural adaptation to climate change in the United States

Malcolm et al. (2012) analyse how crop farmers in the United States will adapt to changing climate conditions and how potential pest pressures and emerging technologies, such as drought-resistant crops, alter the benefits of adaptation. Drought-resistant varieties maintain yields under conditions of reduced precipitation thereby reducing yield losses due to climate change in regions with low precipitation.

The study employed downscaled climate projections from four different general circulation models based on IPCC Special Report on Emission Scenarios (SRES) A1B emission scenario. The Environmental Productivity and Integrated Climate (EPIC) model was used to analyse the impact of each climate scenario on crop yields and the Regional Environment and Agriculture Programming (REAP) model was then used to assess climate-induced changes in regional production patterns and indicators of environmental quality (jointly with EPIC results).

The study focused on the yield-related impacts of increased average temperatures, regional changes in average precipitation, CO_2 effect, increased incidence of pests and commodity price changes. However, neither the impact of extreme weather events nor the potential for expanding irrigation area water use were addressed.

The impacts of climate change vary widely across regions mainly due to changes in the direction and magnitude of precipitation. Farmers are able to reduce the impact by altering cultivated crops, crop rotations and production practices and redistribution of production across regions help to alleviate the impact on national commodity markets.

National acreage changes are relatively small (from 0.2 to 1%) when farmers adapt while regional changes can be larger. For most commodities adaptation helps to dampen the price rises, although corn and soybean prices increase due to lower national yields. Net returns to crop producers are estimated to be on range of USD 3.6 billion increase and USD 1.5 billion decrease depending on the climate change scenario. Pest damage may decrease the net returns by USD 1.5 billion to USD 3.0 billion. Impact on commodity prices vary widely from decline of wheat prices to potential increase of soybean (from -4% to 22%) and corn prices (from -2% to 6%) depending on climate scenario. As regards environmental effects it has been estimated that due to cropland increase nitrogen losses increase by 1.4%-5.0% and soil erosion range from -0.9% to 1.2%.

The introduction of drought-tolerant varieties increased yields by 10%–15% in drier but non-irrigated area, reduced total planted acreage across all the climate scenarios, increased economic returns nationally and in regions that cultivate them, and reduced prices for corn, soybeans, wheat and cotton.

Source: Malcom, S., E. Marshall, M. Aillery, P. Heisey, M. Livingston, and K. Day-Rubenstein (2012), "Agricultural Adaptation to a Changing Climate – Economic and Environmental Implications Vary by US Region", ERR-136, U.S. Department of Agriculture, Economic Research Service, July 2012.

Knowledge gaps and surprises: Pests, diseases and weeds, pollinators

In spite of these efforts towards assessing the costs and benefits of adaptation options for agricultural water management, it is important to recognise that remaining knowledge gaps can bias the results. Perhaps a major example of knowledge gaps related to water and agriculture is the influence of pests, diseases and weeds, which biological cycles are heavily dependent on weather conditions, in particular water parameters such as rainfall, moisture, etc. After all, in light of an agronomic perspective, a cultivated field can be seen as a place where several species compete for available resources. Hence if changes in the water cycle affect crop growth, it also affects the growing conditions of pests, diseases and weeds. How climate change will affect the relative competitiveness between crops and weeds? What kind of agricultural management practices would be able to adapt cropping systems to this new equilibrium? There is less evidence in the literature about these impacts on crop production, but it is already well known that even today, pests and diseases are still responsible for substantial yield losses — pests, pathogens and weeds are estimated to be responsible of, respectively 18%, 16% and 34% of crop yield losses (Walthall et al., 2012). A proper analysis of the impact of climate change should consider the functioning of the cultivated field as a competition between several organisms, all of them being expected to be affected by changes in the pattern of weather variables due to climate change.

Box 2.4. The IMPACT modelling approach

The impact modelling approach combines several modules including a hydrology model, a water model, and a GTAP economic model in order to simulate the impacts of different climate forcing scenarios on water demand and supply, rainfed and irrigated areas, food prices, trade, population, labour and capital stocks, and labour productivity (**Figure 2.2**). Recent applications of this integrated modelling approach include for instance sub-Saharan Africa, where IMPACT allowed analysing scenarios such as expanded irrigation areas and increased productivity in agriculture.

Figure 2.2. Structure of the IMPACT model

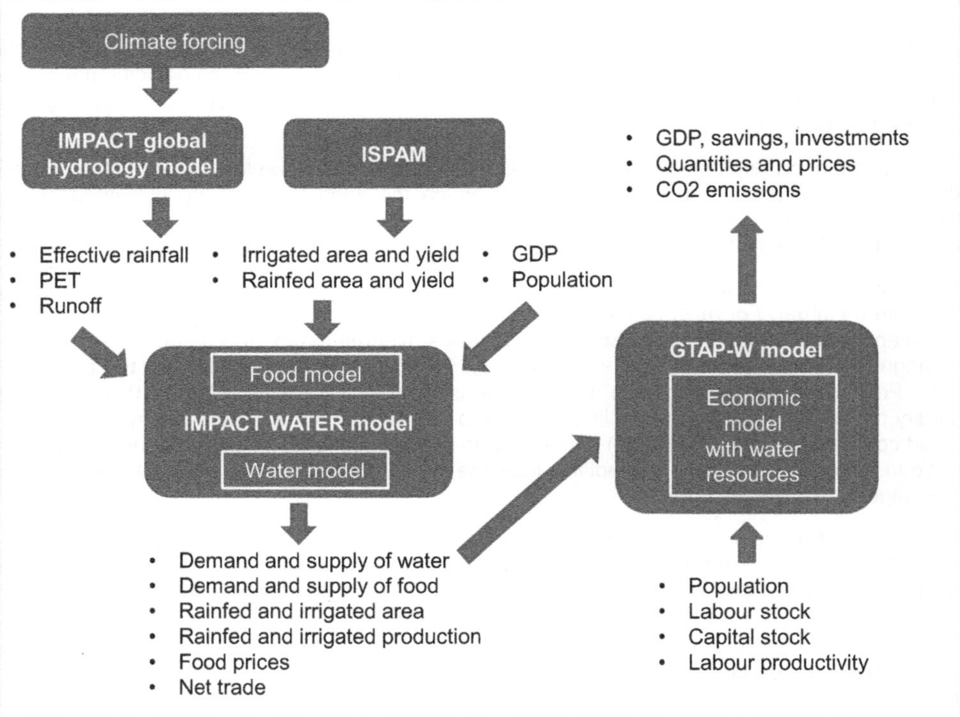

Source: Calzadilla A., T. Zhu, K. Rehdanz, R.S.J. Tol and C. Ringler (2012), "Economy-wide Impacts of Climate Change on Agriculture – Case Study for Adaptation Strategies in Sub-Saharan Africa", in A. Dinar and R. Mendelsohn (Ed.), *Handbook on Climate Change and Agriculture,* Edward Elgar Publishing, Cheltenham, United Kingdom.

Public policies fostering on-farm adaptation of water management

Most of the adaptation strategies described above can be considered as private production decisions, and as such do not call for specific public policy intervention. There may be rationale for government intervention to foster adoption of these on-farm adaptation practices in some cases.

A first aspect is the switching costs of adjustments that can be high for some farmers. In some regions, the impacts of climate change on the water cycle and other weather variables may be so substantial that farmers may have to completely rethink their farming systems – in some cases even relocate or exit farming. In that case, they may not be able to bear the financial cost of doing so. Public policy can thus play a role to smooth adjustment costs related to these deep structural adjustments, while ensuring at the same time that new land uses remain compatible with environmental objectives. Research and development policies and more generally innovation systems, could contribute to foster the adjustment process. In cases farms are financially unable to adapt, targeted adaptation funds or temporary technological subsidies could facilitate the adaptation process, and avoid situations where farms are locked into non-adapted production systems. These temporary interventions should be limited in time and well-targeted in order to avoid any distortive impact on production systems and production choices.

In cases the impacts of climate change are not too substantial for farms, or at least can be managed by a limited set of adjustments in agricultural management practices, there may still be room for public policy intervention to create an enabling environment towards adaptation. This can take the form of information provision, technical assistance and experience sharing of best adaptation practices. This can incite risk-averse farmers to adopt new cropping and livestock systems with uncertain results. Another rationale for such intervention lies in behavioural economics, which shows the importance of non-financial incentives in decisions, as well as the potentially high efficiency-cost ratio of "nudging" incentives (see OECD, 2012a for a recent survey in the farm sector). More generally, education, training and investment in skills should be a key component of successful on-farm adaptation strategies, for both structural and marginal adjustments.

When fostering on-farm adaptation through education, innovation policies and technical assistance, a specific role for governments is also to take care of environmental consequences of these interventions, so that private and social incentives coincide as far as possible. This is the subject of the next section dedicated to water management policies for adaptation at the watershed level.

Water management policies for adaptation at the watershed level

Agriculture water withdrawals in the OECD represent a major share of total freshwater withdrawals – 44% on average in the OECD area – as well as a source of environmental externalities related to water systems such as nutrient pollution, pesticides and soil erosion (OECD, 2010a; OECD, 2012b; OECD, 2013c). This has important implications for the adaptation of water systems at the watershed level, both in terms of water allocation between agriculture and other water uses and of environmental externalities. Water has often the characteristics of a common pool resource; hence adaptation of water management in agriculture has implications for other water users. Agricultural management practices can also affect risks of floods in certain watersheds, as agricultural lands can either play the role of pathway of water flows or receptors of water in times of floods. Due to these strong linkages between agriculture and other sectors and users through the water system, adaptation of agricultural water management has a collective dimension, and cannot be limited to a private, farm level issue.

Regardless of the climate challenge, current water policies in agriculture include a set of regulatory rules, economic instruments such as water pricing and quota trading, and collective water rules (OECD, 2010a). The new question that comes with climate change and adaptation is the following one: will the current arrangements – or systems – be able to cope with non-stationary climate conditions? In other words, what are the best policy approaches for rendering water management systems sensitive to climate change issues, both in terms of average changes in climate conditions but also in terms of increased risks of extreme events such as floods and droughts?

Challenges and priorities can vary across countries and regional circumstances. A recent study from FAO (2011) identified for the different regions of the world the expected impacts of climate change on water management in major agricultural systems, related vulnerabilities and possible response options. A summary of their analysis is provided in **Table 2.2**.

Adapting agricultural water management to climate change requires the appropriate signals of water scarcity to be sent to water users. These signals can take several forms in practice. They can consist in regulatory constraints, such as a mandatory limitation or interdiction of freshwater withdrawals in cases of drought, or allocating farmers a right for a maximum volume of water for the whole growing season. Other approaches can be based on economic instruments, such as water pricing and water trading. In this case, the signal of water scarcity takes the form of a price. Economic instruments are well known for their desirable properties in terms of economic efficiency. In practice, assigning well-defined property rights related to water uses is, however, a difficult task, due to the specific nature of water as a good, characterised by its mobility and by usually complex hydrological interactions across water compartments (OECD, 2010a).

The primary objective of water pricing and water trading is not to adapt to climate change, but to increase the efficiency of water allocation. However, the flexible and decentralised characteristics of economic instruments may also be desirable properties in a context of non-stationary, uncertain climate. In the context of climate change, it seems important to distinguish two types of incentives:

- ***Short term incentives***, which reflects the cost of water scarcity within the growing season.

- ***Medium and long-term incentives***, that reflect the real *average* cost of water scarcity in the near future and in the long run.

Short-term incentives to reallocate water during the growing season. If investments related to water efficiency are long run, the issue of water allocation within the growing season is also important aspect of adaptation, as climate change is projected to increase the frequency and severity of extreme weather events. In case of drought, the marginal value of water becomes large for irrigated farms. Within the farm, famers can allocate water on a priority basis to the crops that bring the highest returns. If farmers have different marginal values of water, *ex post* water trading can reallocate water to the most efficient uses, which can mitigate the overall economic losses arising from water shortages and thus increase social welfare.

Structural adaptation requires long run, time consistent incentives. Putting an appropriate price signal that reflects the state of water supply for water users can be considered as a non-regret option, as it aims to improve the efficiency of water management. In the context of adaptation, a signal that reflects the present state of water scarcity is not sufficient. Decisions related to irrigation, water efficiency, switching crop rotations and cropping systems involve in most cases mid to long-term investments, based on farmers' expectations about future economic and climate conditions. For example, if water rights allocated to farmers are well tailored to the present state of water supply, but not dependent on total freshwater available, farmers will form their expectations without taking into account the risk of rising water scarcity.

The flexibility associated with the allocation of water rights is important for the purposes of adaptation. As climate change is a continuous process, water management systems should be able to continuously adapt to new climate conditions. An example of an evaluation of the potential benefits of water markets under climate change is Luo et al. (2010), who model a water trading scheme in the Swift Current watershed in **Canada**. They show that water trading is able to reduce significantly water withdrawals without reducing farm revenues, although the extent of such benefits heavily depend on projected climate conditions, in particular the frequency of dry conditions. Another example is the **United States**, where

Libecap (2011) has identified significant potential gains from water trade between farms, and even higher values for trading between the farm sector and urban users.

Table 2.2. Typology of climate change impacts on water management in major agricultural systems

System	Current status	Climate change drivers	Response options
SNOWMELT SYSTEMS			
Indus system	Highly developed, water scarcity emerging. Sediments and salinity constraints	Twenty years increasing flows followed by substantial reductions in surface water and groundwater recharge. Changed seasonality of runoff and peak flows. Increased flooding. Increased salinity.	Increased water storage and drainage; improved reservoir operations
Northern China	Extreme water scarcity and high productivity		Change in crop and land use; improved soil management
Colorado	Water scarcity, salinity		
HUMID TROPICS			
Rice: Southern China	Conjunctive use of surface and groundwater	Increased rainfall and rainfall variability, with more frequent droughts and floods	Increased storage for second and third season; drought and flood insurances; crop diversification
Rice: Northern Australia	Fragile ecology		
TEMPERATE AREAS			
Northern Europe	High value agriculture and pasture	Increased rainfall; longer growing seasons; increased productivity	Potential for new development Storage development; drainage
Northern America	Cereal cropping; groundwater irrigation	Reduced runoff; increased water stress	Increase productivity Limited options for storage
MEDITERRANEAN			
Southern Europe (Italy, Greece, Spain)	Problems of water scarcity and shortage	Significantly lower rainfall and higher temperatures, increased water stress, decreased runoff	Localised irrigation; transfer to other sectors
Northern Africa	High water scarcity		Localised irrigation; use of groundwater

Source: Adapted from FAO (2011), *Climate change, water and food security*, FAO Water Report No. 36.

Box 2.5. Impediments to water markets formation

- Incomplete understanding of the science of water resource and ecosystem linkages.

- Lack of physical interconnecting networks between water delivery systems supplying agriculture, urban, industrial and other users.

- Uncertainty about the supply and demand for water at a given point in the future.

- Poorly defined property rights, including problems of separating land from water entitlements.

- Defining, securing and agreeing among stakeholders the quantity of water needed in a water basin to sustain environmental values.

- High transaction costs in creating water markets.

- Issues of equity in that water markets are perceived to ignore the poor, and that they are also largely considered to focus on economic efficiency overlooking environmental and social considerations.

- In many circumstances irrigators do not have the opportunity to trade their water entitlements with other users as no markets exist to do so.

Source: OECD (2010a), *Sustainable Management of Water Resources in Agriculture*, OECD Studies on Water, OECD Publishing. doi: 10.1787/9789264083578-en.

Box 2.6. Two examples of flexible re-allocation mechanisms

Volumetric management of water resources in France

In France, water pricing for irrigation mainly covers operations and maintenance costs, and investment costs related to irrigation infrastructure and is not used to reflect scarcity costs (OECD, 2010a). Irrigators have to pay a fee to the Water basin agency, which can be differentiated between areas or irrigation techniques, but its average amount is usually considered as insufficient as an incentive device to reduce water demand by irrigators (Lefebvre and Thoyer, 2012, Erdlenbruch et al., 2013). These fees are used to finance accompanying actions for water-related projects.

One example of water quantity management practice for structural and short-run water shortages is volumetric management, which is applied in some French watersheds such as Charente and Beauce. Under this system, each farmer is allocated, for the timeframe of the irrigation season, a maximum water withdrawal right in volume terms. This global volume is then allocated across time periods, typically weeks or ten-day. Volumes are estimated by a metering device at the farm level, and can be controlled randomly by the Water Police at the end of each period (Lefebvre and Thoyer, 2012). The 2006 water legislation introduces the possibility to issue water withdrawal authorisation to a single organisation responsible for establishing water sharing across irrigators every year. In areas of structural water deficits, the prefect, who is the State's representative in regions or departments, can mandate the creation of such single organisation. In those areas, the single organisation will fully replace the current system based on individual authorisations by 2015. Such evolution is expected to improve stewardship incentives of irrigator, encourage equitable sharing of water and foster collective and local approaches for water management.

In times of water shortages, specific re-allocation rules apply which can be described as follows. The level of water availability is regularly measured by the regional offices of the French ministries of environment and of agriculture. Flows of rivers or groundwater levels are measured at different points of the water system, and compared to some predefined triggering alert thresholds, *étiage* flow trigger (EFT), and the crisis flow trigger (CFT). When water flow is higher than the EFT, there is no restriction for irrigation. If the flow or the water level exceeds the predefined threshold, then restrictions apply either through reduction of the reference volume, or through the prohibition to irrigate for a given number of days in the week or ten-day period (Erdlenbruch et al., 2013). On the basis of these set of pre-defined rules, the decision process involves stakeholders in the framework of a drought committee. These different restriction rules are anyway temporary, and can be revised in light of the evolution of the state of water flow, with the aim of ensuring a balanced water resource sharing between the different uses such as agriculture, industry, tap water and minimum environmental flows.

Water markets in Australia

Australia has been experiencing frequent and severe drought episodes in the last decade, and has undertaken significant reforms of water management policies, especially in the agricultural sector. The specific feature of the Australian system of agricultural water management is the use of water markets.

> Farmers are allocated water entitlements each year, which can be revised according to the state of water reserves and rainfall. There are two distinct water markets: a long-run water market, and a short-run water market. Each market is targeted to a specific timeframe. Short-run water markets allow farmers temporary water exchanges, allowing for a more efficient allocation of water within the growing season, and so a reduction of the overall cost for farmers as whole.
>
> The interesting specificity of the Australian water markets is that individual rights are defined as a share of total volume of water available for a given period of time. This allows the system to continuously, and almost automatically adapt to changing weather and hydrological circumstances in the course of the growing season. Public buybacks allow preserving the integrity of environmental flows. A second feature of the system is that these shares can be exchanged between farmers, in the course of the growing season. These two features taken together allow reducing significantly the overall cost of droughts for the farm sector on the whole, as shown by Mallawaarachchi and Foster (2009) in a study covering the 2007-2008 water shortage experienced by the country.

Although the flexibility and the incentive effects of economic instruments make them potentially interesting policy instruments for adaptation, it seems also important to keep a holistic approach and recognise the complexity of water management. Real world examples of water markets are still limited: **Australia** and **United States** being the two main examples. Several reasons may explain this limited development (see **Box 2.5** from OECD, 2010a). Water markets also require a good identification of pre-existing water rights, an accurate determination of the volumes that can be abstracted and cannot by themselves solve problems of over-allocation. These limitations taken together could explain to some extent a certain degree of path dependency of water management rules (Libecap, 2011), and which itself is an important issue for adaptation of water management to climate change. **Box 2.6** illustrates the relative roles of short-run and long-run incentives in the case of Australia and France.

Developing economic incentives for water management, as any policy reform, can also redistribute already allocated— implicit or explicit — water rights across water users. Hence, if this may result in overall social efficiency gain, there can be winners and losers with the new system. The dependence to institutional path in climate adaptation has for example been raised by Libecap (2011) in the case of **United States**, and should be taken into account to facilitate reforms of water management systems.

When transaction costs impede the development of water trading, water management can be based on a set of *water sharing arrangements* within the farm sector, or between farmers and other water users. These collective water sharing arrangements can be evaluated according to several aspects: efficiency, equity and robustness to changing conditions. This robustness is a crucial aspect of adaptation: will these sharing rules remain stable in times of an increase in the frequency of drought or water demand?

Recent trends in OECD countries are encouraging, although there is still room for improvement. Water policy reforms over the last two decades have witnessed the development of regulatory and economic instruments, notably water pricing (**Table 2.3**). Even if most of the times they do not cover the cost of water scarcity, they already provide incentives for a better management of water resources in agriculture. Recent data from OECD agri-environmental indicators on water resources can shed some light on this question. On average in the OECD zone, agricultural freshwater withdrawal decreased by 0.5% per annum between 1998-00 and 2008-10, compared to an increase of 0.2% per annum between 1990-92 and 1998-00 (OECD, 2013c). **Figure 2.3** presents the recent trends of irrigation water application rates for a set of OECD countries. This shows that most of these countries have witnessed a decrease of their irrigation, at a higher rate in the 2000s decade than in the 1990s. Of course, these trends cannot be attributed to climate change only; increases in population, in urban water demand, as well as the rising concern for a more balanced water sharing between human uses and ecosystems are an important part of the story. But still, adaptation would at least be a co-benefit of these evolutions.

Table 2.3. Water cost recovery in agriculture, OECD countries

		Operation and maintenance cost recovery	
		Less than 100%	100%
Investment costs recovery	Less than 100%	Spain, Greece, Hungary, Ireland, Italy, Mexico, Netherlands, Poland, Portugal Switzerland, Turkey, Korea	Australia, Canada, United States, France, Japan
	100%		Austria, Denmark, Finland, New-Zealand, United Kingdom, Sweden

Source: OECD (2010b), *Climate Change and Agriculture: Impacts, Adaptation and Mitigation*, OECD Publishing. doi: 10.1787/9789264086876-en.

Figure 2.3. Irrigation water application rates, OECD countries: 1990-2010

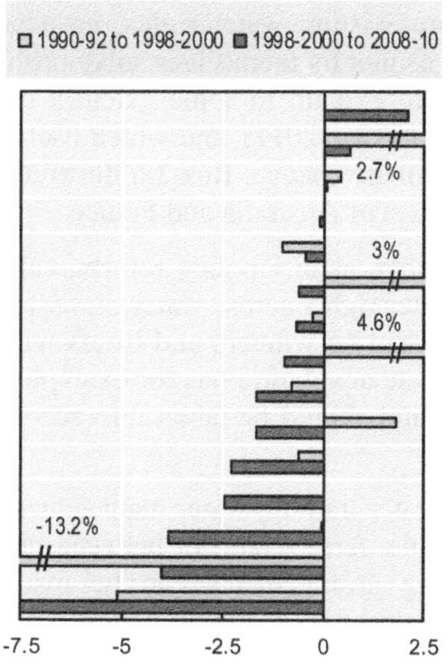

	Irrigation water application rates[1]			Average annual % change	
	Megalitres per hectare of irrigated land			% per annum	
	1990-92[2]	1998-2000[3]	2008-10[4]	1990-92 to 1998-2000	1998-2000 to 2008-10
New Zealand	..	3.4	4.3	..	2.1
Korea	14.3	17.6	18.2	2.7	0.7
Japan	20.6	21.5	21.6	0.4	0.1
Italy	..	7.7	7.6	..	-0.1
Spain	7.0	6.5	6.3	-1.0	-0.4
Israel	5.2	6.6	6.2	3.0	-0.6
Greece	6.3	6.1	5.8	-0.3	-0.7
Turkey	7.9	11.4	10.3	4.6	-0.9
United States	9.1	8.4	7.7	-0.8	-1.7
Mexico	11.4	12.2	10.7	1.8	-1.7
France	3.3	3.1	2.6	-0.6	-2.3
Chile	..	18.1	15.2	..	-2.4
Portugal	10.4	10.4	7.3	0.0	-3.8
Australia	8.7	4.9	3.6	-13.2	-4.0
Denmark	0.9	0.7	0.4	-5.1	-7.5

The figures only include those OECD countries where irrigation area exceeds 5% of total agricultural area, with the exception of Australia where irrigated agriculture is important (irrigation accounts for over 50% of total freshwater withdrawals) but is less than 5% of agricultural land because of the large area under pasture. Countries are ranked in descending order according to average annual % change 1998-00 to 2008-10. Data for Israel refer to agricultural freshwater withdrawals. The statistical data for Israel are supplied by and under the responsibility of the relevant Israeli authorities. The use of such data by the OECD is without prejudice to the status of the Golan Heights, East Jerusalem and Israeli settlements in the West Bank under the terms of international law.

1. Irrigation water application rates are calculated as the quantities of irrigated freshwater withdrawals divided by the irrigated area.

2. Data for 1990-92 average equal to: 1997 for Australia; 1990-91 average for Denmark; 1990 for France, Japan, Korea, Portugal and United States; 1990-92 average for Greece; 1994-95 average for Mexico; 1991 for Spain.

3. Data for 1998-00 average equal to: 2001 for Australia; 1999 for Chile; 1995-96 average for Denmark; 2000 for France; 2000-02 average for Greece; 1998 for Italy; 2000-01 average for Japan; 1997-98 average for Korea; 1998-00 for Mexico, Spain and Turkey; 2006 for New Zealand; and 2000 for Portugal and United States.

4. Data for 2008-10 average equal to: 2007-09 average for Greece; 2006 for Chile; 2002-04 average for Denmark; 2007 for France; 2006-08 average for Israel, Japan and Mexico; 2002-03 average for Korea; 2010 for New Zealand; 2009 for Italy and Portugal; 2005-07 for Spain; and 2008 for United States.

Source: OECD (2013c), *OECD Compendium of Agri-environmental Indicators*, OECD Publishing. doi: 10.1787/9789264186217-en; OECD Agri-environmental Indicators Questionnaires.

Risk management approaches for adapting to increasing risks of droughts and floods

Increasing weather risk due to climate change is expected to lead to a direct reduction of farmers' income in absence of adaptation measures dedicated to reducing the sensitivity of agricultural outcomes to weather risks. Moreover, climate change may not just increase weather risks, but also the level of *ambiguity* about these risks, due to the fact that climate is becoming *non-stationary*, so past experiences are no longer relevant to assess future risks (OECD, 2013b). Economic theory shows that ambiguity tends to reinforce risk aversion, and so the associated welfare impacts on farmers. The overall impact depends of course of adaptation responses, and increasing risk and ambiguity are also incentives for farmers to invest in risk management strategies. The role of risk as a driver of farmers' choices is still debated, but as it is increasing, one can reasonably expect that risk itself, not just expected outcome, becomes a more and more important driver of farmers' choices.

Insurance and compensation policies against droughts and floods

The management of risks related to water – floods and droughts – can take the form of risk sharing arrangements and insurance markets. These solutions can in theory allow the farmers to share risks within the sector, or more broadly with the rest of the economy. In theory, insurance can be an interesting adaptation tool for two reasons.

- It redistributes the burden of climate risks on individual farms by sharing it across farms and other sectors, in a way that – according to economic theory – maximises social welfare.

- By putting a fair price on risks – as long as asymmetric information does not prevent it – it can play the role of an *economic incentive* guiding farms towards efficient, from a social welfare point of view, production choices. For example, if a given crop is becoming especially risky in terms of yields because of climate change, this would be reflected by an increase in the insurance premium for this crop that would directly incite the farmer to reconsider his or her crop allocation.

Such benefits essentially build on the premise that agricultural insurance markets are well-functioning and cover the major sources of risks, which requires certain conditions: the presence of a real risk; the possibility to assess it; statistically independent risks so that the law of large numbers apply. Other major obstacles for the development of private insurance markets could be the presence of fat-tail risks and tail dependency between risks, making insurance portfolios at risk of huge losses and complicating the calculation of insurance premiums and requirements of buffer capital (Kousky and Cook, 2009). Several limitations do in fact already hinder the large development of a private, non-subsidised agricultural insurance markets in the OECD countries, even without considering the issue of climate change. Some risks are already covered, in particular those for which there is some degree of statistical independence across farms or regions – for example, risks of hail. But major risks such as droughts are less well or not at all covered by private insurance markets, at least without government support in most OECD countries (OECD, 2011). Hence, considering carefully the relative advantages and limitations of current agricultural insurance schemes is an important step for fostering insurance as an efficient adaptation tool to climate change.

Perhaps the most specific issue with insurance as an adaptation tool is the non-stationarity of climate, which renders the task of actuaries – to measure risks and put a price on them – more difficult. In practice, it is possible to gradually adjust premiums to reflect the trend followed by weather risks under a changing climate. This allows for a continuous update of the cost of the risk, which can provide valuable guidance to farmers' adaptation decisions (Weinberg, 2012). Nevertheless, climate change also creates some specific challenges for insurance markets. Under non-stationary climate, the frontier is continuously moving from the world of risk – where probabilities can be assessed – to the world of uncertainty – where they cannot be. Insurance firms traditionally cover risks, not uncertainty. Hence to be able to assist agriculture as an efficient adaptation tool, the insurance sector also needs to adapt to these new conditions. With climate change, risks are not just following a trend, they also become more ambiguous. This affects both the demand and supply of insurance. **Figure 2.4** illustrates the expected consequences of an increase in ambiguity for a given risk. In theory, risk increases the demand for insurance (ambiguity reinforces risk aversion), but also the cost of supplying insurance – due notably to costs of capital necessary to deal with large losses.

Figure 2.4. Hypothetical impact of ambiguity on the demand and supply of insurance

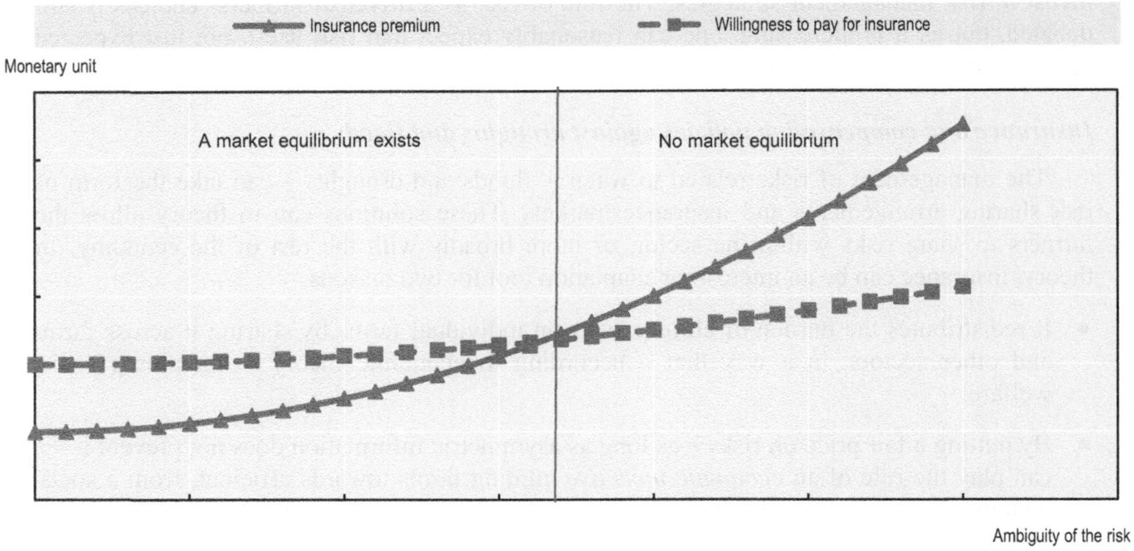

Innovative risk-sharing tools

Recent decades have witnessed the development of alternative instruments for risk sharing and transferring. In particular, weather index insurance consists in basing insurance indemnities on the occurrence of a weather event, which is defined by one, or a set of weather variables. For example, indemnities are triggered when temperature in a given area exceeds a certain level. The principle underlying this instrument is that individual losses are correlated with the weather index. When individual risks are sufficiently correlated in a given geographical area, farmers are covered against the systemic risk related to the weather event. Weather index insurance contracts have several desirable properties: they can allow eliminating transaction costs related to insurance monitoring of individual losses, as well as risks related to moral hazard. Moreover, there is a potential for transferring the underlying risk to financial markets, which can broaden the scope of risk sharing, and thus lead to efficiency gains in terms of collective risk sharing. Weather index insurance has also some limitations: it may be difficult to price; by nature it allows to insure against the systemic component of the weather risk. Hence each individual farmer whose yield is not perfectly correlated with the weather risk would have to retain the idiosyncratic component of the risk.

Several experiences of weather index insurance contracts are in place around the world (Hellmuth et al., 2009). In a context of rising risk of extreme events, even if strong uncertainty remains a major limitation for risk sharing in a non-stationary climate, it is for sure an important dimension of adaptation to look for the most cost-efficient insurance options to increase the resilience of agriculture to climate shocks. There is potential room for these innovative risk management tools to reduce implementation costs of the current insurance and compensation systems. However, rather than a "one size fits all" solution to agricultural insurance, they should be considered as a valuable innovation in a broader set of risk management instruments. The nature, characteristics and relative importance of weather risks vary a great deal across countries, so there is a need for tailored solutions.

Sharing country experiences can play a role in improving agricultural insurance systems. The June 2012 Los Cabos declaration of G20 leaders, under the heading of enhancing food security, endorsed the initiative to create the Platform for Agricultural Risk Management (PARM). The G20 development group meeting in February 2013 endorsed the conceptual note on PARM drafted in collaboration between several international organisations (including OECD) and development agencies. The approach emphasises risk assessment and a holistic approach. PARM will be hosted by the International Fund for Agriculture Development (IFAD) and is expected to start operating in 2014. The objective of PARM is to strengthen Agricultural Risk management in developing countries in a holistic manner and on a demand driven basis. The Platform is thought to facilitate matching between agricultural risk management needs and existing tools, sharing the experiences of different organisations and practitioners. The PARM will be working with a variety of development institutions and international organisations including IFAD, FAO, WFP, OECD and the World Bank.

The need for a holistic approach for managing weather risks under climate change

Risk management against drought and flood risks should not be limited to insurance and compensation policies, but rather be considered in a more holistic perspective (OECD, 2009). On-farm adaptation options can modify the average outcome of farm systems, as well as the risk profile. For instance, irrigation can allow for both increases in average yields and a decrease in yield risk due to rainfall fluctuations. In some cases there can be a trade-off between risk and expected outcome, as shown previously in the case of drought-resistant seeds. Considering in combination the impacts of climate change on both mean yields and yield variability can influence the choice of adaptation priorities.[2] A recent study by González-Zeas et al. (2013) analyses this question in the case of Europe. In this study the combined effects of changes in the mean and overall variability in regional crop productivity are assessed. This provides a possibility to prioritise the agro-climatic regions where climate change impacts must be addressed more immediately. Four cases are identified to resolve adaptation priorities (**Table 2.4**). For example, if yield variability decreases and the mean yield increases, the focus should be on removing barriers to potential opportunities, but if both decrease, adaptation should focus on the average impacts. González-Zeas et al. (2013) used scenario A2 for the period 2071-2100 and compared these to the control scenario for the period 1961-1990. **Figure 2.5** shows changes in mean yields and yield variability in different regions under scenario A2.

As can be seen from the results, the projected climate change is expected to increase crop yields in the boreal, continental south and alpine regions. Mean yields are expected to decrease in other regions. Yield variability increases under the scenario A2 in the continental north, Atlantic central and south, and Mediterranean north and south. Interestingly, in the Mediterranean north and in Atlantic central the changes in yield variability are much larger than those in the mean yields. Adaptation needs and priorities can thus be directly associated with the joint behaviour of both the mean and variability of yields, as illustrated by González-Zeas et al. (2013) in **Table 2.5**.

Another important issue related to the economic cost of extreme events concerns the sharing of risks between farmers and consumers. Due to low price elasticity of most agricultural commodities, production losses due to weather shocks may lead to more than proportional increases in prices, which may compensate in certain cases production losses in terms of farmers' revenues, a phenomenon sometimes called "natural hedging." In these situations, consumers of agricultural commodities face weather risks through price risks.

Trade openness could in principle contribute to a certain extent to reducing price volatility related to extreme weather events, and redistribute weather risks between farmers and consumers.

Table 2.4. Prioritisation of the adaptation requirements

		Mean yield change (%)	
		(-)	(+)
Yield variability change (%)	(+)	Priorities for intervention (impacts and risks management)	Adaptation focus on reducing variability (risk management)
	(-)	Adaptation focus on average impacts	Adaptation focus on eliminating barriers to potential opportunities

Source: Gonzalez-Zeas D, Quiroga S, Iglesias A, Garrote L (2013), "Looking beyond the average agricultural impacts in defining adaptation needs in Europe", *Regional Environmental Change*. doi:10.1007/s10113-012-0388-0.

Figure 2.5. Changes in mean yields and yield variability in different regions under climate change

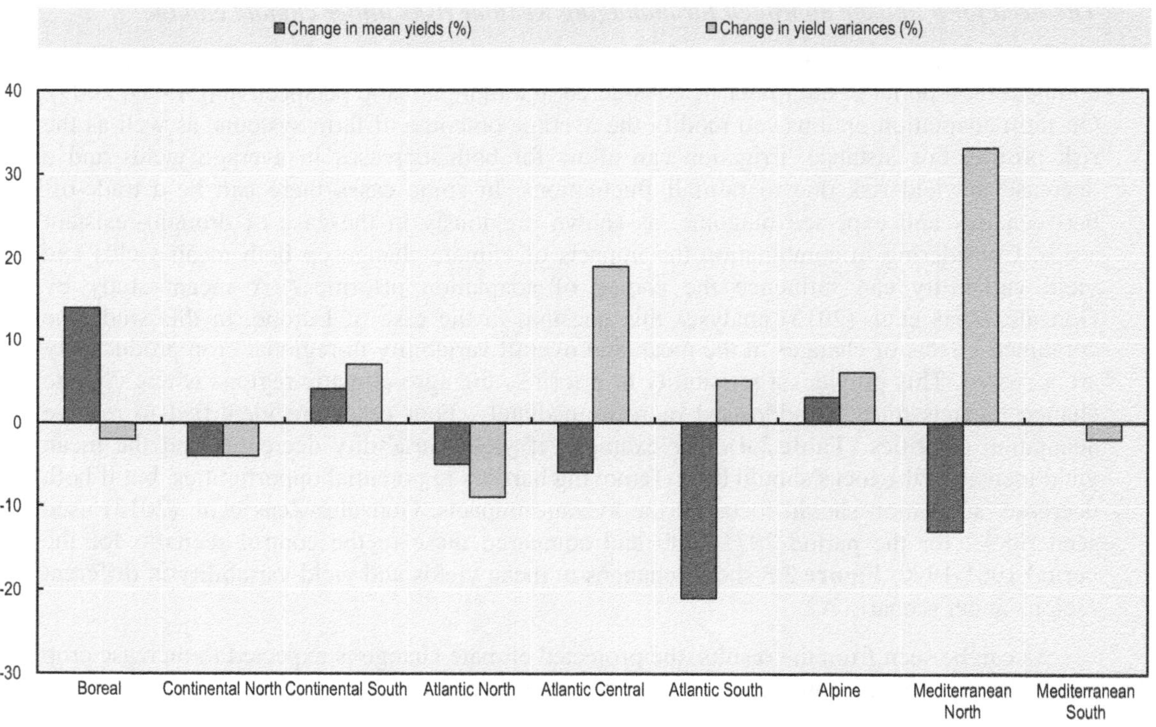

Source: Gonzalez-Zeas D, Quiroga S, Iglesias A, Garrote L (2013), "Looking beyond the average agricultural impacts in defining adaptation needs in Europe", *Regional Environmental Change*. doi:10.1007/s10113-012-0388-0.

Table 2.5. Adaptation strategies that address changes in mean yields and yield variability in European agroclimatic regions

Adaptation needs	Potential adaptation measures
Adaptation focused on average impacts	• Changes in crops and cropping patterns • Changes in cultivation practices • Increased input of agro-chemicals • Introduce new irrigation areas • Develop climate change resilient crops • Diversify livelihood • Relocate farm processing industry
Adaptation focused on reducing variability	• Promote insurance • Provide supplemental irrigation • Shift crops from vulnerable areas • Improve soil moisture retention capacity
Adaptation focused on both changes in the mean and the variability	• Implement regional adaptation plans • Provide advisory services • Promote research on technology and biotechnology • Promote research on water use efficiency • Promote research on management and planning
Adaptation focused on eliminating barriers to potential agriculture opportunities	• Develop adaptation plans to maintain optimal farming conditions and increased crop productivity • Provide advice based on expert judgement

Source: Gonzalez-Zeas D, S. Quiroga, A. Iglesias, L. Garrote (2012), "Looking beyond the average agricultural impacts in defining adaptation needs in Europe", *Regional Environmental Change,* Vol. 12. doi:10.1007/s10113-012-0388-0.

Agricultural policy coherence and the role of market drivers

Market drivers, price signals, and international risk-sharing

Progress has been made throughout OECD area in implementing agricultural policy reforms that have increasingly decoupled farm income support and other payments from farmers' production decisions and thus they have reduced distortions in input and output markets. Through decoupling of support payments from prices and production the government allows market supply and demand conditions as well as market prices better influence farmers' production decisions. This could contribute positively to climate change adaptation since farmers would not need to continue cultivation of only program crops but they could diversify cropping patterns on the basis of supply and demand conditions. Decoupling also contributes to reduced environmental pressures and reduced trade distortions.

Because climate change results in new patterns of temperature and rainfall also agricultural production patterns and comparative advantage changes with consequent impacts on commodity trade flows. Trade allows comparative advantage to be fully employed while restrictions on trade flows risk worsening the economic damage from climate change. International trade moderates and transmits climate change impacts on commodity markets and in the case of adaptation trade pools the risk since yield losses in one region can be offset through imports (Nelson et al., 2009). Changing trade flows are important mechanism to offset – at least partially – the negative productivity effects of climate change.

Nonlinear effects of temperatures and an increase in the frequency of extreme events like floods are likely to change both the average outcome and variability of yields with climate change (Rosenzweig et al., 2002; Urban et al., 2012). Competitive markets for commodities are likely to incorporate changing expectations about both the level and volatility of yield, which should influence futures prices as well as option prices.

In a competitive market for storage of grains and oilseeds, higher volatility and perceptions of shifting future yield trends can influence prices and storage of commodities along the transition path to a warmer world. Preliminary research looking at these kinds of adaptive storage responses to climate change suggest that increased volatility of food production could be partly buffered by increased competitive storage, such that prices are less likely to become more variable even if year-to-year commodity production becomes more variable (Tran et al., 2012). International buffer stock mechanisms have been widely judged to have only limited success in reducing price volatility, may lead to potentially very high costs, and are vulnerable to speculative attacks (FAO, OECD et al., 2011). However, government intervention could have a potential role in providing the enabling environment to promote private storage and competitive storage markets.

These theoretical market responses are encouraging, but they also might be distorted or circumvented by changing trade policies or export bans. Recent experience with abrupt changes in trade policies, like the Russian Federation's recent ban on wheat exports after extreme heat and fire induced severe crop losses in 2010, could upset natural market forces that might otherwise temper price volatility in a changing climate.

These issues are closely related to the food security debate. Major extreme water events such as floods and droughts and water scarcity could reduce average agricultural production, and raise production risk, and so price risk. Ensuring stable food supply would thus require securing agricultural water management, in particular in large producer countries. The recent OECD report on *Global food security: Challenges for the food and agriculture system* (2013a) underlines the need to consider climate change as an important part of the food security issue.

Policy sequencing: Fostering an enabling policy environment for adaptation

Conventional policy design wisdom is that one should first remove policy failures, such as environmentally harmful input subsidies, that exacerbate environmental market failures, and only after that address any remaining environmental market failures with targeted and tailored policy interventions. Typical example of policy failure is irrigation water subsidy, which leads to overuse of scarce water resources and nutrient and pesticide leaching and thus exacerbates environmental damage. Similarly market price support and output subsidies provide incentives to increase environmentally harmful input use, such as fertiliser, pesticide and irrigation water. Moreover, these types of coupled policies also increase opportunity costs of environmental policies and partly offset their environmental benefits. Furthermore, farmers' income problems need not to be addressed with traditional "broad brush" agricultural policy instruments as income objectives could also be met by general social and welfare policies (OECD, 2002; 2008). Following this general advice on policy sequencing governments could consider modifying those policies that may lead to maladaptation, such as government financed or subsidised insurance, emergency relief and farm income programs which allow farmers to continue production in high-risk low productivity locations (Mendelsohn, 2010).[3]

Interactions between mitigation and adaptation of agricultural water management

Adaptation and mitigation are sometimes closely linked with potential trade-offs and synergies. Naturally, the stronger the current mitigation policies are, the less the need will be for adaptation in the long run. On the one hand agricultural mitigation practices that increase nutrient and water retention and prevent soil degradation can increase resilience to droughts and flooding. On the other hand adaptation measures that reduce tillage, increase crop rotations and promote green cover can contribute to mitigation efforts (OECD, 2010b).

Most agricultural GHG mitigation and soil carbon sequestration practices, such as adoption of no-till or green fallowing, have complex site-specific water resource and water quality effects. For example, the adoption of no-till is likely to reduce sediment and nutrient runoff, but may increase herbicide runoff. Changes in land use, such as conversion of cropland to green fallow, may have these effects and also affect water resources. If these co-effects are significant, then they should be explicitly addressed when designing policies to mitigate GHG emissions and sequester carbon. However, this may come at greater implementation costs.

Moreover, the spatial variation of these co-effects may be large, and policy design and implementation may need to reflect heterogeneous supply of both mitigation and sequestration capacity and co-effects through targeting and tailoring of climate policy incentives. Naturally, this holds true as regards the GHG co-effects of water quality, water resource and biodiversity conservation and sustainable use policies.

Numerous agricultural mitigation measures, their emission abatement potential and abatement costs have been analysed in the literature (for a comprehensive global review see Smith et al., 2008). Recently, Pellerin et al. (2013) analysed abatement potentials and costs of ten technical mitigation measures in France and concluded that agriculture sector has significant abatement potential without affecting significantly production systems. Mitigation activities in agriculture can be classified in four main categories as follows (McCarl and Schneider, 2000; Smith et al., 2007; Baker et al., 2013):

- *Reducing emissions* by changing crop and livestock management, switching land allocation between crops, and changing land use from crops to green fallow or forests.

- *Enhancing absorption of atmospheric carbon* by creating or expanding carbon sequestered in sinks. This largely involves changes in tillage intensity, land use, and afforestation (Lal, 2004; Murray et al., 2005).

- *Providing products which substitute for GHG emission intensive products* like fossil fuels or building materials in turn displacing emissions from those sources (McCarl, 2008).

- *Developing technical advances that can reduce GHG emissions* in absolute terms or per unit of output produced (Baker et al., 2013).

Because they can require changes in cropping systems, most mitigation activities available in agriculture can have – directly or indirectly – potential impacts on water resources and water quality. For example:

- In terms of *water quantity*, widespread development of bioenergy feedstocks may cause additional water use and, in some places, lead to irrigation water use expansion.[4] Crops like sugar cane replacing grasses or other crops may use additional water due to their larger evapotranspiration requirements. On the other hand, moving from crops to grass may not only decrease net long term GHG emissions but may also increase infiltration of water into ground water reservoirs and may also slow down runoff shifting more to subsurface flows that are released to streams slower making water available over a longer time period.

- In terms of *water quality*, a number of mitigation related strategies like manure management, fertiliser management, reduced tillage and land conversion from crops to grass or forest have long been recommended practices to reduce nitrogen, phosphorus, pesticide, and sediment runoff and leaching into surface and ground waters with accompanying water quality effects. More generally, water quality effects can occur when mitigation practices alter soil erosion rates, fertiliser and pesticide input uses, and amount and nutrient content of animal manure in turn altering sediment, nutrient and pesticide runoff. Both ground and surface water can be affected.

Understanding the linkages between mitigation and water requires taking into account not only the direct effects of the mitigation activity on water systems, but also its influence on other agricultural management practices which can, in turn, have impacts on water systems. For example, farmer's adoption of no-till farming in order to increase soil carbon sequestration affects not only tillage practices but also type and amount of applied inputs, such as fertiliser and pesticide, which in turn have impacts on water quality.

Characterising the effects of climate change mitigation activities on water – in both quantity and quality terms – requires a set of environmental indicators related to water. Below is a set of the main environmental pressures on water systems that can be altered – directly and indirectly – by mitigation practices:

- *Agricultural water withdrawal* can be affected by refocusing water conservation policies and institutions to encourage broader integration of improved irrigation application systems with both on-farm and watershed-level water management.

- *Agricultural water runoff* can be affected by mitigation activities such as conversion of agricultural land from cultivated crops to grassland or forest, and creation of wetlands where water runoff is largely reduced.

- *Nutrient and pesticide runoff from crop production* can be affected by mitigation activities such as changes in fertiliser and pesticide type and use, adoption of reduced tillage and no-till, and conversion of agricultural land from cultivated crops to green fallow or forests.

- *Nutrient runoff from livestock and manure application* can be affected by mitigation activities such as the reduction of livestock numbers and the improved manure management.

In addition to the linkages described in the framework presented above, a number of mitigation policies in agriculture may have even more indirect effects on water quantity and quality; this involves a phenomenon which in the Kyoto protocol language has typically been called *leakage* (Murray, McCarl and Lee, 2004). This refers to the fact that, given that a mitigation strategy leads to a reduction in commodity production – directly reducing yield per hectare or reducing the hectares farmed – this diversion in turn puts pressure on the marketplace to replace the lost production. In turn this stimulates more production on lands not directly within the mitigation project. For example, the first generation biofuel products that use conventional crops like corn or sugarcane as feedstocks reduce the amount of corn and sugarcane in the marketplace and causes producers in other regions to expand their corn and sugarcane production. This production expansion can come about through either intensification or extensification.

Due to leakage, producers in other regions may pursue actions towards intensification such as adopting irrigation, increasing fertiliser and pesticide application, with potential environmental pressures on water systems. Intensification can also increase sediment runoff and in turn diminish water quality (Baker et al., 2010). Pfeiffer and Lin (2010, 2013) provide examples where improved irrigation technology reduces per acre water use and production costs in turn causing an expansion in land area farmed, with on the whole a potential increase in total water use.

The extensification reaction involves land use change. In particular diminished production in one area can lead to increases in production somewhere else which in turn diminish off farm water quantity and reduce water quality when the new acreage uses inputs like fertiliser and pesticides. There is also an indirect effect on livestock where increased commodity prices and thus increased feed costs cause reduction in the livestock herd, as argued in Murray et al. (2005) and Ugarte et al. (2008) and accompanying improved water quality due to less manure application and related nutrient runoff.

Mitigation practices can be grouped into large classes like land use change, crop management, livestock management, bioenergy, and technological progress. Next these classes and some major practices that fall underneath them are briefly discussed. Practices examined here are meant to illustrate linkages between selected mitigation practices and water. As such they do not represent a comprehensive list of all available mitigation options in the agriculture sector.

Table 2.6 summarises the main linkages between mitigation practices and water resources and quality. In this Table, the + sign means that mitigation activity improves water quality or quantity situation and - denotes the worsening of the situation. +/- means that overall impact is not generally determined, but would depend on specific case. A more extensive review of mitigation practices can be found in Annex C.

The main challenge for policy making in integrating adaptation and mitigation strategies is implementation costs, due to the very site specific nature of the interactions, and the cost if information. Some policies aim at supporting a set of agricultural management practices that include a package of co-benefits on carbon sequestration, water and biodiversity. This approach allows for sure limiting implementation costs, but the question of the optimal levels of regulation and fine tuning of economic instruments such as taxes and subsidies remains. This debate is beyond the case of water and carbon, but includes all potential positive and negative externalities arising from agricultural production.

The linkages between mitigation and water presented in this section should be considered as rather general because of the regional specificity of water quantity and quality effects and the inherent uncertainty even at local scales as well as because in some cases, overall effects are ambiguous.

Most of the agricultural mitigation practices analysed here are practices that have been implemented for other agronomic, economic or environmental goals such as yield enhancement, water conservation, input cost reduction, or addressing some other environmental objectives, such as water quality and biodiversity. These practices include, for example, reduced tillage and no-till, manure and fertiliser management, and precision irrigation. Thus, much of the discussion in this section is based on the results of these potential greenhouse gas mitigating practices as they have been implemented in other settings. While it is anticipated that the practices will largely have the same implications in a greenhouse gas mitigation setting, there may be alterations in the means of implementation that could have some implications on water quantity and quality.

Table 2.6. Summary linkages by mitigation activity on water resources and quality

		Water runoff	Nutrient and pesticide runoff and leaching	Nutrient runoff from livestock manure	Fossil fuel use	Irrigation water withdrawal
Direct land use change	Crops to pasture	+	+		+	+
	Crops to forest	+	+		+	+
	Marginal and pasture to crops and bioenergy	-	-		-	-
Agricultural management — Cropping management	Tillage change and landscape contouring	+	+/-		+	+
	Crop mix and perennials	+	+			+/-
	Irrigation management	+	+		+	+
	Fertiliser and nutrient management		+		+	
	Other chemical use reduction		+			
Agricultural management — Animals	Manure management			+		
	Breeding and animal species choice		+	+/-		+
Bioenergy — Energy form and process	Liquid fuels			+/-	+	
	Electricity		+		+	
	Pyrolysis - biochar	+	+		+	
	Conventional crops and their residues	+/-	+/-		+	
	Energy crops	+/-	+		+	
	Animal wastes			+	+	
	Processing by-products		+		+	

+ means that water quality or quantity situation improves; - denotes the worsening of the situation; +/- means that overall impact is not determined, but would depend on specific case.

The linkages presented in the table need to be considered as very general because of the regional specificity of water quantity and quality effects and the inherent uncertainty even at local scales as well as because in some cases, overall effects are ambiguous even in qualitative terms.

Sources: McCarl, B.A., and U.A. Schneider (2000), "US Agriculture's Role in a Greenhouse Gas Emission Mitigation World: An Economic Perspective", *Review of Agricultural Economics*, Vol. 22(1); Oleson, J. and J.R. Porter (2009), "Deliverable 10: Adaptation and Mitigation", PICCMAT project report, European Commission, Brussels; Smith, P., et al. (2007), "*Agriculture*" In B. Metz *et al. (eds.), Climate Change 2007: Mitigation. Contribution of Working Group III to the Fourth Assessment Report of the Intergovernmental Panel on Climate Change*, Cambridge University Press, United Kingdom and New York, NY, USA.

The analysis on the effects of mitigation practices on water quantity and quality needs to be considered in general qualitative terms and not quantitatively. This is because the literature illustrates the spatial heterogeneity of effects due to differences in soil types, topography, climate and many other factors. In particular, there has been a lot of literature on how reduced intensity of tillage increases soil carbon sequestration (Lal et al., 1998). However, Blanco-Canqui and Lal (2008) find cases where this is not the case and states "the idea that no-tillage would also enhance soil organic carbon sequestration as an additional benefit of no-tillage technology needs a careful re-examination". Similarly, studies of the nitrogen removal capacity of buffer strips show that their effectiveness varies widely dependent on soil type, subsurface hydrology, and subsurface biogeochemistry (Mayer et al., 2007).

In some cases overall effects of mitigation practices can be ambiguous. For example, changes in crop rotation may at the same time affect the amount of fertilisers, thus the overall effect on nutrient balances may be difficult to assess or be dependent on local conditions.

Summary

Adaptation strategies for agricultural water management need to accommodate considerable uncertainty about the impacts of climate change on agricultural systems, and the costs and benefits of adaptation options. This uncertainty, however, does not call for inaction. In many OECD countries, water supply is expected to decrease and become more volatile, while water demand is projected to increase. There is thus room for immediate action in favour of improving water use efficiency in agriculture as a no-regrets strategy.

On-farm adaptation such as adopting drought-resistant crop varieties, improving irrigation techniques and switching to less vulnerable and more resilient cropping and livestock systems have some potential to reduce the vulnerability of farms to climate change. Government policies targeted to innovation, education and soft technical measures can play a role in facilitating the natural trend towards on-farm adaptation.

Adaptation of agricultural water management to climate change needs regulatory and economic instruments that allow revealing the social price of water. Recent progress in that direction during the last decades has allowed improving irrigation water use efficiency in agriculture and among other water users.

Because of projected increases in risks of extreme weather events such as droughts and floods, adaptation strategies will have to consider insurance and compensation systems as part of the tools able to increase the resilience of agricultural systems. These tools should be able to deal not just with increasing risks, but also increasing uncertainty about these risks. Innovative risk management tools such as weather index insurance may be able to play a role, although it remains important to have a holistic approach of risk management.

Open trade is an important vehicle to fully reflect shifting comparative advantage due to climate change while also pooling the risk so that yield losses in a given region can be offset through imports. Similarly adaptive storage can help to buffer against increased volatility of commodity production and prices.

Synergies and trade-offs between climate change mitigation and adaptation policies need to be explicitly addressed in policy design to fully benefit from complementarities and to minimise the risk of conflicts.

In this chapter, there has been no attempt to prioritise climate change adaptation and mitigation options, since climate change impacts are highly uncertain, site-specific, and dependent on the socio-economic context. Generally, the choice of adaptation policies and their sequencing need to resort to decision-making approaches that are robust to deep uncertainty, on a case-by-case basis. Moreover, since climate change adaptation is a dynamic process, adaptation policies need to allow for flexibility and dynamic learning.

Notes

1. Since Knight (1921), the distinction is usually made between *risks*, which probabilities and consequences can be assessed quantitatively, and *uncertainty*, which refers to situations in which they cannot. This distinction is helpful but needs to be qualified case by case: most of the time risks also include some degree of uncertainty. *Deep uncertainty*, as analysed by Hallegatte et al. adds to Knightian uncertainty two dimensions: the fact that there could be disagreements between equally valid views of the world; and the fact that decisions to adapt are dependent over time (*cf.* glossary).

2. Focussing on the mean and variance is interesting as a first approach; however, it does not fully reflect the complexity of the risks. Weather risks are known to have typically non-normal probability distributions, such as fat-tailed, which would require using complementary criteria such as higher order statistical moments.

3. It is nevertheless important to analyse policy failures on a case-by-case basis to avoid possible adverse effects of policy reforms. For example, irrigation can in some watersheds contribute to instream flows and groundwater recharge due to the hydrological characteristics of the water system. On a more general level, this is the problem of a second-best policy solution.

4. In certain OECD member countries and the European Union, sustainability criteria have the objective to avoid these unintended environmental consequences.

References

Baker, J.S., B.A. McCarl, B.C. Murray, S.K. Rose, R.J. Alig, D.M. Adams, G.S. Latta, R.H. Beach, and A. Daigneault (2010), "Net Farm Income and Land Use under a U.S. Greenhouse Gas Cap-and-Trade", *AAEA Policy Issues*, Issue 7.

Baker, J.S., B.C. Murray, B.A. McCarl, S.J. Feng, and R. Johansson (2013), "Implications of Alternative Agricultural Productivity Growth Assumptions on Land Management, Greenhouse Gas Emissions, and Mitigation Potential", *American Journal of Agricultural Economics*, Vol. 95, pp. 435-441.

Barnett, T.P., J.C. Adam and D.P. Lettenmaier (2005), "Potential impacts of a warming climate on water availability in snow-dominated regions", *Nature*, Vol. 438, No. 7066, pp. 303–309.

Berry, S.T., M. J. Roberts and W. Schlenker (2012), "Corn production shocks in 2012 and beyond: Implications for food price volatility", In *The Economics of Food Price Volatility*, University of Chicago Press.

Blanco-Canqui, H. and R. Lal (2008), "No-Tillage and Soil-Profile Carbon Sequestration: An On-Farm Assessment", *Soil Science Society of America Journal*, Vol. 72, No. 3, pp. 693-701.

Brisson, N., P. Gate, D. Gouache, G. Charmet, F.-X. Oury and F. Huard (2010), "Why are wheat yields stagnating in Europe? A comprehensive data analysis for France", *Field Crops Research*, Vol. 119, Issue 1, pp. 201-212.

Butler, E.E. and Huybers, P. (2013), "Adaptation of us maize to temperature variations", *Nature Climate Change*, Vol. 3, pp. 68–72.

Calzadilla A., T. Zhu, K. Rehdanz, R.S.J. Tol and C. Ringler (2012), "Economy-wide Impacts of Climate Change on Agriculture – Case Study for Adaptation Strategies in Sub-Saharan Africa", in A. Dinar and R. Mendelsohn (Ed.), *Handbook on Climate Change and Agriculture*, Edward Elgar Publishing Ltd., Cheltenham, UK.

Cimato, F. and M. Mullan (2010), "Adapting to Climate Change: Analysing the Role of Government", *Defra Evidence and Analysis Series*, Paper 1.

Dauphiné A., Provitolo D. (2007), "La résilience : un concept pour la gestion des risques", *Annales de géographie*, Vol. 2 (654), pp. 115-125.

Dixit, A. and R. Pindyck (1994), *Investment Under Uncertainty*, Princeton University Press.

EEA (2012), "Climate change, impacts and vulnerability in Europe 2012 – An indicator-based report", European Environment Agency.

Erdlenbruch, K., S. Loubier, M. Montginoul, S. Morardet, M. Lefebvre (2013), "La gestion du manque d'eau structurel et des sécheresses en France", Sciences, Eau et Territoire, No. 11, p. 78-85

FAO (2011), "Climate change, water and food security", *FAO Water Report* No. 36.

FAO, OECD et al. (2011), "Price Volatility in Food and Agricultural Markets: Policy Responses", Inter-Agency Report to the G20 including contributions by FAO, IFAD, IMF, OECD, UNCTAD, WFP, the World Bank, the WTO, IFPRI and the UN HLTF.

Fleischer A. and Pradeep Kurukulasuriya (2012), "Reducing the Impact of Global Climate Change on Agriculture – the Use of Endogenous Irrigation and Protected Agriculture Technology", in Handbook of Climate Change and Agriculture, chapter 16, pp. 355-381.

Fleurbaey, M. and Zuber, S. (2012), "Climate policies deserve a negative discount rate", Fondation Maison des Sciences de l'Homme, WP-2012-19.

Gitz,V. and A. Meybeck (2012), "Risks, Vulnerabilities and Resilience in a Context of Climate Change", in Meybeck, A., J. Lankoski, S. Redfern, N. Azzu and V. Gitz (eds.), *Building* resilience for adaptation to climate change in the agriculture sector – Proceedings of a Joint FAO/OECD Workshop, Rome, 23-24 April 2012.

Gollier, C. (2013), *Pricing the Planet's Future:The Economics of Discounting in an Uncertain World*, Princeton University Press.

Gonzalez-Zeas D, S. Quiroga, A. Iglesias, L. Garrote (2013), "Looking beyond the average agricultural impacts in defining adaptation needs in Europe", Regional Environmental Change, Vol. 12.

Hallegatte S., F. Lecocq and C. de Perthuis (2011), "Designing Climate Change Adaptation Policies. An Economic Framework," *Policy Research Working Paper* No. 5568, World Bank, Washington DC.

Hallegatte, S., A. Shah, R. Lempert, C. Brown, S. Gill (2012), "Investment Decision Making Under Deep Uncertainty – Application to Climate Change", *Policy Research Working Paper* No. 6193, World Bank, Washington DC.

Hellmuth M.E., D.E. Osgood, U. Hess, A. Moorhead and H. Bhojwani (eds) (2009), *Index insurance and climate risk: Prospects for development and disaster management*, Climate and Society No. 2. International Research Institute for Climate and Society (IRI), Columbia University, New York, USA.

IPCC (2012), Managing the Risks of Extreme Events and Disasters to Advance Climate Change Adaptation. A Special Report of Working Groups I and II of the Intergovernmental Panel on Climate Change, Cambridge University Press, Cambridge, UK, and New York, NY, USA, pp. 582.

Johansson, P.-O. (1991), *An introduction to modern welfare economics*, Cambridge University Press, Cambridge, United Kingdom.

Knight, F.H. (1921) *Risk, Uncertainty, and Profit*, Boston, MA: Hart, Schaffner & Marx; Houghton Mifflin Company.

Kousky, C. and M. Cooke (2009), "Climate Change and Risk Management – Challenges for Insurance, Adaptation, and Loss Estimation", Resources for the Future Discussion Paper 09-03.

Lal, R., J.M. Kimble, R. Follett and C.V. Cole. (1998), *The Potential of U.S. Cropland to Sequester Carbon and Mitigate the Greenhouse Effect*. Sleeping Bear Press, Chelsea, MI, pp. 128.

Lefebvre, M. and S. Thoyer (2012), "Risque sécheresse et gestion de l'eau agricole en France", in Études et Synthèses, N°2012-02, UMR LAMETA.

Libecap, Gary D. (2011), "Institutional Path Dependence in Climate Adaptation: Coman's 'Some Unsettled Problems of Irrigation'", *American Economic Review*, 101(1), pp. 64-80.

Luo, B., I. Maqsood and Y. Gong (2010), "Modeling climate change impacts on water trading", *Science of the Total Environment*, Vol. 408, Issue 9, pp. 2034-2041.

Malcom, S., Marshall, E., Aillery, M., Heisey, P., Livingston, M. and K. Day-Rubenstein (2012), "Agricultural Adaptation to a Changing Climate – Economic and Environmental Implications Vary by U.S. Region", ERR-136, US Department of Agriculture, Economic Research Service, July 2012.

Mallawaarachchi T. and A. Foster (2009), "Dealing with irrigation drought: the role of water trading in adapting to water shortages in 2007–08 in the southern Murray–Darling Basin", ABARE Research Report 09.6 to the Department of the Environment, Water, Heritage and the Arts, Canberra.

Mayer, P.M., S.K. Reynolds, M. D. McCutchen and T. J. Canfield (2007), "Meta-Analysis of Nitrogen Removal in Riparian Buffers", *Journal of Environmental Quality*, Vol. 36, No. 4, pp. 1172-1180.

McCarl, B.A. (2008), "Bioenergy in a greenhouse gas mitigating world", *Choices*, 23(1), pp. 31-33

McCarl, B.A., and U.A. Schneider (2000), "US Agriculture's Role in a Greenhouse Gas Emission Mitigation World: An Economic Perspective", *Review of Agricultural Economics*, Vol. 22(1), pp. 134-159.

Mendelsohn, R. (2010), *Agriculture and Economic Adaptation to Climate Change*, COM/TAD/CA/ENV/EPOC(2010)40/FINAL.

Murray, B.C., A.J. Sommer, B. Depro, B.L. Sohngen, B.A. McCarl, D. Gillig, B. de Angelo and K. Andrasko (2005), *Greenhouse Gas Mitigation Potential in US Forestry and Agriculture*, EPA Report 430-R-05-006, November.

Murray, B.C., B.A. McCarl and H-C. Lee (2004), "Estimating Leakage from Forest Carbon Sequestration Programs", *Land Economics*, Vol. 80(1), pp. 109-124.

Nelson, G.C. (2009), *Climate change: Impact on agriculture and costs of adaptation.* International Food Policy Research Institute.

OECD (2013a), *Global Food Security: Challenges for the Food and Agricultural System*, OECD Publishing.
doi: 10.1787/9789264195363-en

OECD (2013b), *Water and Climate Change Adaptation: Policies to Navigate Uncharted Waters*, OECD Studies on Water, OECD Publishing, Paris.
doi: 10.1787/9789264200449-en.

OECD (2013c), *OECD Compendium of Agri-environmental Indicators*, OECD Publishing, Paris.
doi: 10.1787/9789264186217-en.

OECD (2012a), *Farmer Behaviour, Agricultural Management and Climate Change*, OECD Publishing, Paris.
doi: 10.1787/9789264167650-en.

OECD (2012b), *Water Quality and Agriculture: Meeting the Policy Challenge*, OECD Studies on Water, OECD Publishing, Paris.
doi: 10.1787/9789264168060-en.

OECD (2011), *Managing Risk in Agriculture: Policy Assessment and Design*, OECD Publishing, Paris.
doi: 10.1787/9789264116146-en.

OECD (2010a), *Sustainable Management of Water Resources in Agriculture*, OECD Studies on Water, OECD Publishing, Paris.
doi: 10.1787/9789264083578-en.

OECD (2010b), *Climate Change and Agriculture: Impacts, Adaptation and Mitigation*, OECD Publishing, Paris.
doi: 10.1787/9789264086876-en.

OECD (2008), "Agricultural Policy Design and Implementation: A Synthesis", *OECD Food, Agriculture and Fisheries Papers*, No. 7, OECD Publishing, Paris.
doi: 10.1787/243786286663.

OECD (2006), *Cost-Benefit Analysis and the Environment: Recent Developments*, OECD Publishing, Paris.
doi: 10.1787/9789264010055-en.

OECD (2002), *Agricultural policies in OECD countries. A positive reform agenda*, OECD Publishing, Paris.
doi: http://dx.doi.org/10.1787/9789264199682-en

Ortiz-Bobea (2012), "Understanding Heat and Moisture Interactions in the Economics of Climate Change Impacts and Adaptation on Agriculture", Working paper.

Ortiz-Bobea, A. and R.E. Just (2012), "Modeling the structure of adaptation in climate change impact assessment", *American Journal of Agricultural Economics*, published online.

Pellerin S., L. Bamière, D. Angers, F. Béline, M. Benoît, J.P. Butault, C. Chenu, C. Colnenne-David, S. De Cara, N. Delame, M. Doreau, P. Dupraz, P. Faverdin, F. Garcia-Launay,

M. Hassouna, C. Hénault, M.H. Jeuffroy, K. Klumpp, A. Metay, D. Moran, S. Recous, E. Samson, I. Savini and L. Pardon (2013), *Quelle contribution de l'agriculture française à la réduction des émissions de gaz à effet de serre ? Potentiel d'atténuation et coût de dix actions techniques*. Synthèse du rapport d'étude, INRA (France), 92 p.

Pfeiffer, L., and C.-Y.C. Lin (2013), "Does Efficient Irrigation Technology Lead to Reduced Groundwater EXtraction? Empirical Evidence", Unpublished paper, University of California, Davis.

Pfeiffer, L. and C.-Y.C. Lin (2010), "The effect of irrigation technology on groundwater use". *Choices*, 25 (3).

Pindyck, R. (1989), "Irreversibility, uncertainty, and investment". Policy Research Working Paper Series No 294, The World Bank.

Rosenzweig, C., F.N. Tubiello, R. Goldberg, E. Mills and J. Bloomfield (2002), "Increased crop damage in the us from excess precipitation under climate change", *Global Environmental Change* Vol. 12, No. 3, pp. 197–202.

Saleth, R.M., Ariel Dinar and J. Aaprius Frisbie (2011), "Climate change, drought and agriculture: the role of effective institutions and infrastructure", in *Handbook on Climate Change and Agriculture,* Ariel Dinar and Robert Mendelsohn.

Seo, S.N. and R.Mendelsohn (2008), "An analysis of crop choice: Adapting to Climate Change in South American Farms", *Ecological Economics,* Vol. 67, No. 1, pp. 109–116.

Sinclair, T.R. and R.C. Muchow (2001), "System analysis of plant traits to increase grain yield on limited water supplies", *Agronomy Journal*, Vol. 93, No. 2, pp. 263–270.

Smit, B., I. Burton, R.J.T. Klein and J. Wandel (2000), "An Anatomy of Adaptation to Climate Change and Variability", *Climatic Change* Vol. 45, pp. 223–251.

Smith, P., D. Martino, Z. Cai, D. Gwary, H. Janzen, P. Kumar, B. McCarl, S. Ogle, F. O'Mara, C. Rice, B. Scholes, O. Sirotenko, M. Howden, T. McAllister, G. Pan, V. Romanenkov, U. Schneider and S. Towprayoon (2007), "Policy and technological constraints to implementation of greenhouse gas mitigation options in agriculture", *Agriculture, Ecosystems and Environment 118*, 6-28.

Smith, P., D. Martino, Z. Cai, D. Gwary, H. Janzen, P. Kumar, B. McCarl, S. Ogle, F. O'Mara, C. Rice, B. Scholes, O. Sirotenko, M. Howden, T. McAllister, G. Pan, V. Romanenkov, U. Schneider, S. Towprayoon, M. Wattenbach and J. Smith (2008), "Greenhouse gas mitigation in agriculture", *Philosophical Transactions of the Royal Society, Biological Science* N°363, 789-813.

Tran, A.N., J.R. Welch, D. Lobell, M.J. Roberts and W. Schlenker (2012), "Commodity prices and volatility in response to anticipated climate change", In 2012 Annual Meeting, August 12-14, 2012, Seattle, Washington, No. 124827, Agricultural and Applied Economics Association.

Ugarte, D., He, K.L. Jensen, B. C. English and K. Willis (2008), "Estimating Agricultural Impacts of Expanded Ethanol Production: Policy Implications for Water Demand and Quality", Paper presented at American Agricultural Economics Association Annual Meeting, Orlando, FL, July 27-29.

Urban, D., M.J. Roberts, W. Schlenker and D.B. Lobell (2012), "Projected tem- perature changes indicate significant increase in interannual variability of us maize yields", *Climatic change, Vol.112, No. 2,* pp. 1–9.

Walthall, C.L. et al. (2012), Climate Change and Agriculture in the United States: Effects and Adaptation, USDA Technical Bulletin 1935, Washington, D.C., pp. 186.

Weinberg, M. (2012), "Agricultural response to a changing climate: the role of economics and policy in the United States of America", in Meybeck, A., J. Lankoski, S. Redfern, N. Azzu, and V. Gitz (eds.) *Building resilience for adaptation to climate change in the agriculture sector – Proceedings of a Joint FAO/OECD Workshop*, Rome, 23-24 April, 2012.

Chapter 3.

Conclusion and key policy implications

This chapter presents key policy recommendations to build more resilient agricultural and water systems in face of the challenges posed by climate change.

Recognise that water is central to future adaptation to climate change. Although there are substantial knowledge gaps on the impacts of climate change on the water cycle, in particular at the local level, it is expected that the availability of water will change significantly and will be more volatile in the next decades in many countries. Some countries may face drier conditions, others wetter conditions. Most will have to deal with a changing, constantly evolving "water landscape."

As far as possible, fill knowledge gaps at the scale where adaptation takes place in order to foster it and reduce the risk of maladaptation. Water is a core input of agricultural production, so it is crucial for adaptation planning to improve our knowledge of the impacts of climate change on agricultural production. In particular,

- Better understand the specific role of water, in combination with other factors – temperature, CO_2, etc. – in agricultural production can help the agricultural sector to assess the potential benefits of agricultural management practices and innovative cropping and livestock systems.

- Better assess the linkages between climate change, agriculture and water quality.

Because agricultural water management involves public goods, externalities and risk management issues, private adaptation to climate change is not equal to collective adaptation. *A consistent adaptation strategy for adapting agricultural water management thus needs to consider the following five levels of actions* and their linkages.

- On-farm: adaptation of water management practices and cropping and livestock systems;

- Watershed: adaptation of water supply and demand policies in agriculture and with the other water users (urban and industrial) and uses (ecosystems);

- Risk management: adaptation of risk management systems against droughts and floods;

- Agricultural policies and markets: adaptation of existing agricultural policies and markets to the changing climate;

- Interactions between mitigation and adaptation of agricultural water management.

Create an enabling environment for on-farm adaptation. A prerequisite of water adaptation strategies is to create a favourable economic and social environment. Improvement of knowledge on the impact of climate change is part of the enabling environment, but should be completed by other actions, notably:

- Investing in education and training of farmers to climate change and its challenges, in order to improve the adaptive capacity of farmers. As climate change is far in the future and highly uncertain in its local consequences, there is a risk that the problem is not sufficiently considered by farmers given the already full set of challenges that agriculture has to meet today.

- Fostering agronomic and hydrological innovation to foster technical progress allowing to reduce the vulnerability of farm systems to climate change.

- Mixing financial, e.g. through economic instruments, and non-financial incentives. Promotion of collective learning and collective action, sharing knowledge and participation to local adaptation plans can constitute such non-financial incentives, and could also foster the social acceptability of reforming water allocation systems towards more flexibility. Benchmarking between farmers can also increase the rate of diffusion of efficient technical responses to changing climate conditions, e.g. using new communication and information technologies.

- In cases of structural change, public policy could help smooth switching costs across time by adaptation planning and offering temporary financial assistance in clearly defined circumstances.

Adaptation of agricultural water management requires combining the development of *flexible and robust systems of water allocation*, to allow for efficient reallocation of water in a context of strong uncertainty about future water supply and non-stationary climate with a *time consistent, long-run incentive strategy for matching water demand and supply*. Water allocation systems that allow both price and quantity to fluctuate in response to system shocks are desirable both under the existing environment and as a means for providing adaptive capacity with respect to climate change. The path to such more efficient, flexible and robust allocation systems is not always easy and can take time, hence the need to start by now a gradual improvement approach, including the following aspects of the problem:

- Recognising the complexity of water management, characterised by knowledge gaps on the functioning of water systems, on relationships between water compartments, and on relationships between agricultural management practices and water quantity and quality.

- Developing technical equipment, methods and infrastructure to measure water flows and stocks in watersheds. This is a *sine qua non* condition for monitoring impacts of climate change and will provide data for adaptation planning. It will eventually be used as the basis for the development of economic instruments, such as water pricing and trading. Determination of volumes that can be abstracted in a sustainable manner, acknowledging uncertainties resulting from climate change, is also important.

- Establishing water rights, when possible and appropriate. As an alternative, setting up collective arrangements for managing the resource.

- Developing the most efficient mix of instruments: regulatory, economic and collective arrangements.

- Developing the use of economic instruments – water pricing and water markets – which are potentially useful solutions to combine flexibility and long-run incentive, although there are still significant impediments to adoption of such tools for water allocation. As water availability decreases and transaction costs related to these instruments are reduced, there is an increasing room for their development.

- Ensuring a medium to long-run stable horizon to form farmers' expectations and invest in adaptation. Flexibility and short-run efficiency are necessary but not sufficient. This is especially the case in regions where climate change will require substantial, structural adjustment of farm systems and water systems.

- Taking into account the constraints of political economy and distributional impacts of adaptation strategies, which can be major obstacles to the development of more efficient and flexible systems. The inertia of complex systems to adapt, and the weight of past decisions should not be underestimated.

Well-designed risk-sharing arrangements can reduce the burden of increasing risk and uncertainty of weather events such as floods and droughts, and thus contribute to the resilience of agriculture to climate change.

- Catastrophic risks can be managed by combining public-private partnership and innovative risk management tools such as weather index insurance. The costs and benefits of these risk-sharing arrangements should be assessed on a case-by-case basis, as the risk profiles of agricultural systems can widely differ across countries. Such innovative tools can potentially play a role by improving the efficiency of insurance systems, but are still at an experimental stage.

- Clarifying the boundaries between catastrophic and non-catastrophic risks in the design of private and public policy tools. Current frontiers are expected to be moving in a context of non-stationary climate, so a regular update is important for keeping the objective of efficient risk management in the course of climate change.

- In view of policy coherence, the incentive effects of the risk management systems on agricultural production should be carefully considered. Although it can in certain cases be difficult to implement, fair pricing of risks provides farmers a continuous incentive signal to adapt their risk management strategies. Under-pricing of risk, through free of charge *ex post ad hoc* compensation or insurance premium subsidies, can reduce the vulnerability of farms in the short-run, but can provide incentives to increase risk exposure in the longer-run.

- More generally, in addition to specific innovative tools, adopting a holistic risk management perspective, from farm practices to insurance tools, could also contribute to the resilience of the farm systems.

Fostering an enabling policy and market environment for adaptation of agricultural and water systems. Policy and market drivers form the overarching environment within which adaptation strategies take place. Policy failures can increase the cost of adaptation measures. Efforts towards policy coherence are thus required, in particular:

- Removing incentives that are inciting to water overconsumption – e.g. energy subsidies, in link with irrigation water pumping; environmentally distortive production incentives – so they do not offset the simultaneous efforts to promote efficient water pricing; another example are insurance subsidies that incite farmers to increase their risk exposure.

- Fostering free trade, in order to allow the mutual benefits of comparative advantages to be employed. As climate change is expected to redistribute the global map of comparative advantages, this enables flexibility and fair price incentives to reallocate production and adapt agricultural systems.

- Well-functioning competitive commodity storage markets can help farmers and consumers to smooth price shocks due to extreme water events such as droughts and floods across time, reducing the overall cost of increasing volatility of production arising from climate change. Government intervention could have a potential role in providing the enabling environment to promote private storage and competitive storage markets.

Climate change mitigation practices may have positive or negative implications on agricultural water management and on water quality. The potential synergies and trade-offs between mitigation and agricultural management practices are, however, site-specific and for many cases there are substantial knowledge gaps. Although this is a complex matter, it is important to recognise these linkages in the design of mitigation policies, to reduce the risk of conflict between mitigation and water policy objectives, and to maximise potential synergies.

Beyond this set of policy recommendations, it seems important to pursue efforts in research and policy analysis of the linkages between climate change, water and agriculture. Challenges related to the adaptation of agricultural water management to climate change are by nature complex, technical, and context-specific. Some aspects could also have been developed further, such as the adaptation of livestock systems to new water conditions under climate change, the role of agriculture in the management of floods, and the role of research and development and innovation systems. All these aspects would require detailed assessments that are beyond the scope of the present report, which aims at paving the way for a comprehensive and sound policy approach towards resilient agricultural and water systems.

Glossary

Adaptation	Initiatives and measures to reduce the vulnerability of natural and human systems against actual or expected climate change effects. Various types of adaptation exist, e.g. anticipatory and reactive, private and public, and autonomous and planned. Examples are raising river or coastal dikes, the substitution of more temperature-shock resistant plants for sensitive ones, etc. (IPCC, 2007a).
Carbone dioxide fertilisation	The enhancement of the growth of plants as a result of increased atmospheric CO_2 concentration. Depending on their mechanism of photosynthesis, certain types of plants are more sensitive to changes in atmospheric CO_2 concentration (IPCC, 2007a).
Climate model	A numerical representation of the climate system based on the physical, chemical and biological properties of its components, their interactions and feedback processes, and accounting for all or some of its known properties. The climate system can be represented by models of varying complexity, that is, for any one component or combination of components a spectrum or hierarchy of models can be identified, differing in such aspects as the number of spatial dimensions, the extent to which physical, chemical or biological processes are explicitly represented, or the level at which empirical parameterisations are involved. Coupled Atmosphere-Ocean General Circulation Models (AOGCMs) provide a representation of the climate system that is near the most comprehensive end of the spectrum currently available. There is an evolution towards more complex models with interactive chemistry and biology. Climate models are applied as a research tool to study and simulate the climate, and for operational purposes, including monthly, seasonal and inter-annual climate predictions (IPCC, 2007b).
Crop water deficit	The difference between crop evapotranspiration and precipitation that is effective for crop growth.
Crop water requirement	Quantity of water required by a crop in a given period of time for normal growth under field conditions. It consists mainly of evaporative consumption. Usually crop water requirement is expressed in water depth per unit area.
Deep uncertainty	According to Hallegatte et al. (2012), refers to a situation in which the following elements are present: Knightian uncertainty (ignorance of the probabilities and consequences of future possible states of the world); multiple divergent but equally-valid world-views, including criteria of success; decisions which adapt over time and cannot be considered independently.

Evapotranspiration	Water lost to the atmosphere from the ground surface. It is the combined process of evaporation of water stored on and below the ground surface, and the transpiration of groundwater by vegetation.
Extreme weather event	The occurrence of a value of a weather or climate variable above (or below) a threshold value near the upper (or lower) ends of the range of observed values of the variable (IPCC, 2012). A broader and simpler definition includes all weather events such as extreme precipitation, flood, drought, heat wave, hurricane, cyclone, typhoon and tornado.
Maladaptation	Any changes in natural or human systems that inadvertently increase vulnerability to climatic stimuli; an adaptation that does not succeed in reducing vulnerability but increases it instead (IPCC, 2001).
Resilience	The ability of a system and its component parts to anticipate, absorb, accommodate, or recover from the effects of a hazardous event in a timely and efficient manner, including through ensuring the preservation, restoration, or improvement of its essential basic structures and functions (IPCC, 2012).
Resistance	The capacity of a given system to absorb a shock.
Uncertainty	Since Knight (1921), one usually makes the distinction between *risks*, which probabilities and consequences can be assessed quantitatively, and *uncertainty*, which refers to situations in which they cannot.
Vulnerability	The propensity or predisposition of a system to be adversely affected by a shock (IPCC, 2012).

Glossary references

Hallegatte, S., A. Shah, R. Lempert, C. Brown, S. Gill (2012), "Investment Decision Making Under Deep Uncertainty – Application to Climate Change", *Policy Research Working Paper* No. 6193, World Bank, Washington DC.

IPCC (2001), *Climate Change 2001: Synthesis Report. A Contribution of Working Groups I, II, and III to the Third Assessment Report of the Intergovernmental Panel on Climate Change*, Watson, R.T. and the Core Writing Team (eds.), Cambridge University Press, Cambridge, United Kingdom, and New York, NY, United States.

IPCC (2007a), *Climate Change 2007: Impacts, Adaptation and Vulnerability. Contribution of Working Group II to the Fourth Assessment Report of the Intergovernmental Panel on Climate Change*, M.L. Parry, O.F. Canziani, J.P. Palutikof, P.J. van der Linden and C.E. Hanson (eds.), Cambridge University Press, Cambridge, United Kingdom.

IPCC (2007b), *IPCC 2007: Climate Change 2007: The Physical Science Basis. Contribution of Working Group I to the Fourth Assessment*, Report of the Intergovernmental Panel on Climate Change, Cambridge University Press, Cambridge, United Kingdom and New York, NY, United States.

IPCC (2012), Managing the Risks of Extreme Events and Disasters to Advance Climate Change Adaptation. A Special Report of Working Groups I and II of the Intergovernmental Panel on Climate Change, Cambridge University Press, Cambridge, UK, and New York, NY, United States.

Knight, F.H. (1921), *Risk, Uncertainty, and Profit*, Hart, Schaffner & Marx; Houghton Mifflin Company, Boston, MA.

Annex A.

Linkages between climate change, water and agricultural production

Figure A.1. The agricultural production cycle, as impacted by climate change

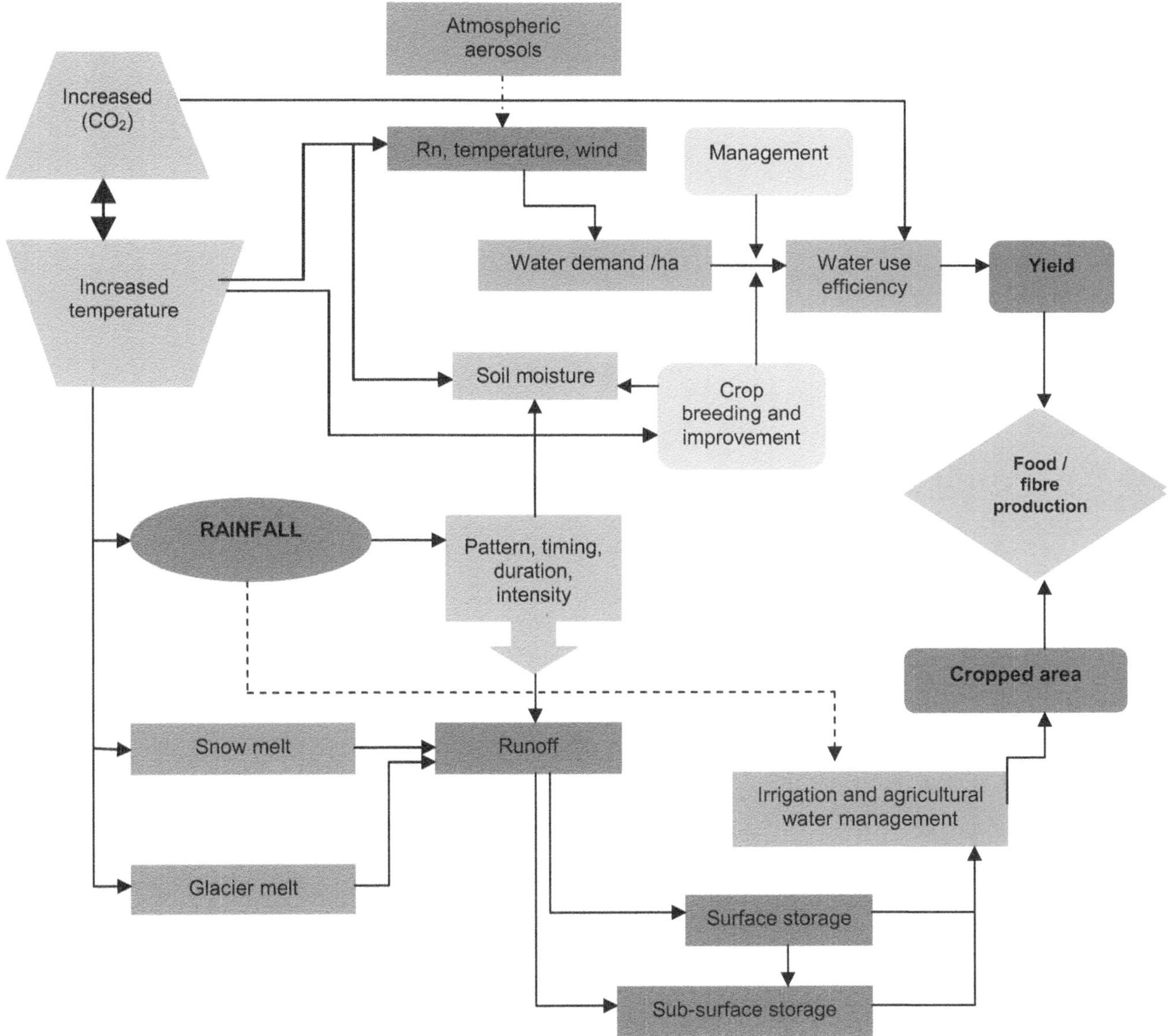

Source: FAO (2011), "Climate change, water and food security", *FAO Water Report*, No. 36, Rome.

Annex B.

Trends in temperature anomalies

Figure B.1. Area of the world covered by temperature anomalies

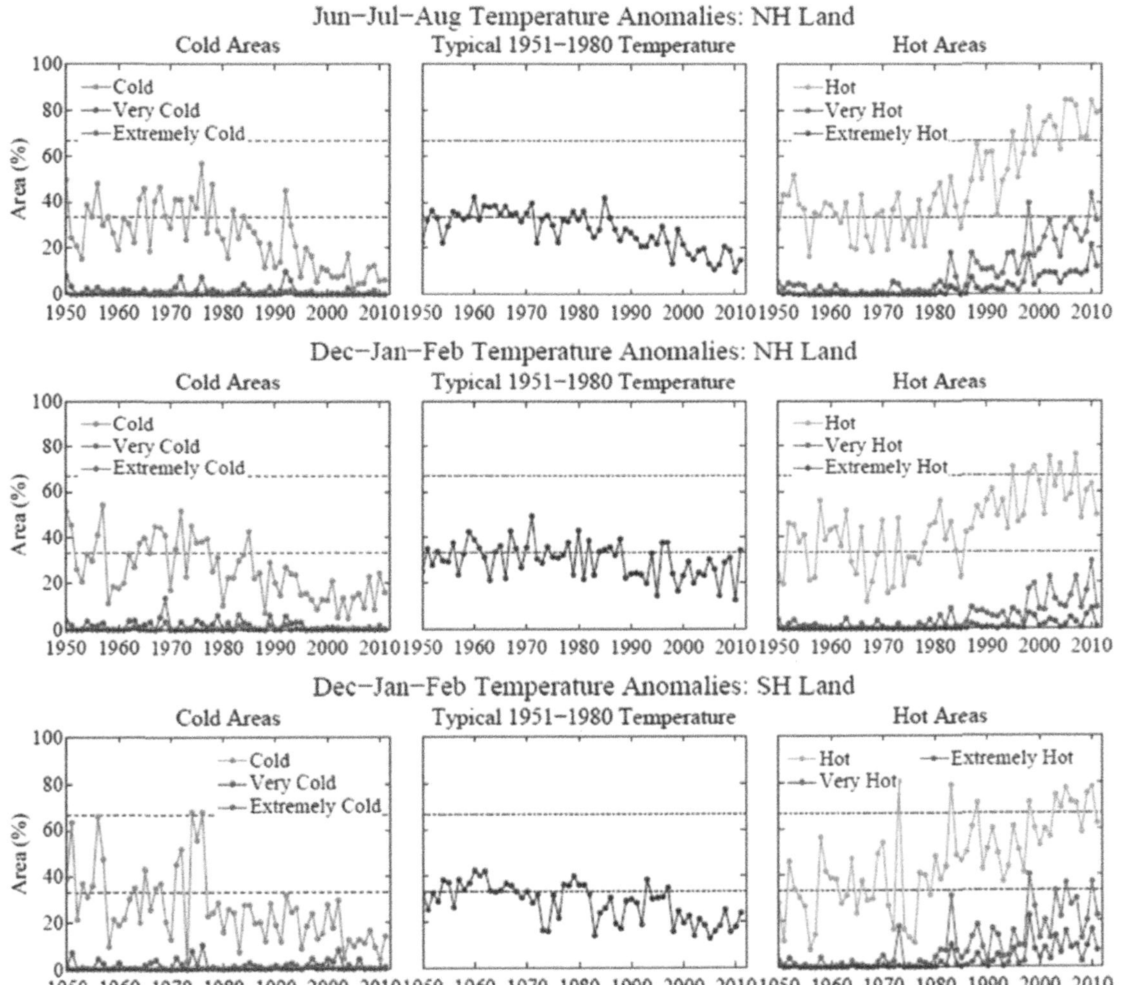

Notes: Categories are defined as hot ($\sigma > 0.43$), very hot ($\sigma > 2$), and extremely hot ($\sigma > 3$), with analogous divisions for cold anomalies. These anomalies are relative to 1951-1980 climatology with σ from the detrended 1981-2010 data, but results are similar for the alternative choices for standard deviation.

Source: Hansen, J., M. Sato, and R. Ruedy (2012), "Perception of climate change", *Proceedings of the National Academy of Sciences of the United States of America*, Vol. 109, 14726-14727, E2415-E2423, doi:10.1073/pnas.1205276109.

Annex C.

A review of the water resource and water quality implications of selected mitigation practices

Direct land use change

The first category of mitigation activities involves land use change where land moves towards more carbon sequestering uses. This includes shifts from crop land to grasslands or forests, and cases where land moves into bioenergy production from croplands, grasslands, forests or wetlands. Individual cases are discussed below.

Crops to pasture

A possible land-use mitigation strategy involves taking land used for crops and moving it into grass. Such a change generally results in reduced soil disturbance, fossil fuel, pesticide, lime and fertiliser inputs which increase soil carbon sequestration and reduce GHG emissions. In terms of water quantity, if crop land was irrigated there would be a reduction in net water use. Furthermore, when land is reestablished as grass it covers the land throughout the year Although more water is used when this occurs, surface water runoff and water infiltration into the underlying groundwater are decreased (Leterme and Mallants, 2011).

There are also strong implications for water quality. Erosion is generally reduced by covering the land in grass, which in turn reduces sediment runoff into the water. Runoff of nitrogen, phosphorus and pesticides are generally reduced because less of these inputs are used (Weller et al., 2003) and because the grass impedes surface flow, thus reducing runoff. If only part of the land is converted to grass and these are placed along waterways (e.g. buffer strips), this tends to filter out erosion, nitrogen and phosphorus that might otherwise runoff into surface waters (Van Dijk et al., 1996; Mayer et al., 2007).

Crops to forest

Another possible land use-based mitigation strategy involves moving cropland into forests. This generally involves reducing soil disturbance and reduced use of nutrients with implications for water quantity and quality. In terms of water quantity, there would be a change in runoff volume with less occurring in forest soils, although because of the vegetation cover more water is generally used than for cropland.

There are also implications for water quality. A number of studies and authors argue that water quality is highest in areas where land use is dominated by forests (Brown and Binkley 1993, Clark et al., 2000, Fulton and West, 2002). Erosion is generally reduced, as well as runoff from nitrogen, phosphorus and pesticides.

Marginal and pasture lands to crops and bioenergy

Another possible land use-based mitigation strategy involves the creation of substitute products by cultivating energy crops on land previously deemed marginal and in grass, brush, trees or some other land cover. This generally involves, at least initially, increasing soil disturbance and the application of pesticide and fertiliser inputs. Such actions can have a number of implications for water quantity and quality. In terms of water quantity, if the energy crop is irrigated then there would be an increase in that category of water use. However, the implications on water depend on prior vegetation with use reduced if the land is moved from forests and perhaps insignificant change if moved from grasslands. A switch from crops depends on relative water use, but Bhardwaj et al. (2011) indicates that an increase is likely.

There are implications for water quality. If the land was degraded and in poor shape with substantial erosion, then planting energy crops may reduce sediment runoff. On the other hand, if the land was originally covered with grass or trees, then erosion would, at least initially, increase and this would in turn increase sediment runoff. Moreover, nitrogen, phosphorus and pesticide runoff would likely be increased because the energy crop would typically involve greater use of fertiliser and pesticides relative to the prior land use (Schnoor et al., 2007).

Agricultural management practices

The second major category of mitigation activities that have water implications involves changes in agricultural management. Such actions are designed to decrease emissions by reducing production inputs and/or to increase soil carbon sequestration. These fall into crop and livestock categories.

Tillage practice

There are number of sequestration enhancing or emission-reducing possibilities that involve changes in cultivation practices and crop mix. Change in tillage practice, such as switching from conventional tillage to no-till, reduces soil disturbance and runoff while also leaving more crop residues on the field. In terms of water quantity, the extra organic matter holds water and may reduce irrigation needs (Holland, 2004). Such an action would reduce the volume of water running off the land, and thus reduce surface water supplies but increase infiltration to groundwater (Pikul and Aase, 2003). In terms of water quality, tillage change has long been advocated as a way to manage erosion (Beasley, 1972; Moldenhauer et al., 1983). It also reduces nutrient runoff (Holland, 2004; Rabotyagov et al., 2010), but may increase herbicide runoff (Wright et al., 2012).

Crop mix and perennials

Another mitigation alternative that reduces emissions and can enhance soil carbon sequestration involves crop mix shifts and potential switches to perennials. Generally, this involves moving from emission intensive crops to those with lower emissions or moving from annual crops to perennials. In terms of water quantity, there may well be a change in irrigation needs as water use varies substantially per hectare and per tonne produced between alternative crops (Mekonnen and Hoekstra, 2010). Use of perennials may reduce runoff and groundwater infiltration levels, as well as sediment runoff. Substantial variations may occur depending on crop mix and crop management. For example, Rabotyagov et al. (2010) show large changes depending on crop mix and management (phosphorus loads varying by a factor of six).

Irrigation management

Improved irrigation management, when combined with improved irrigation efficiency and integrated within larger watershed-scale institutional water management tools can improve both water resource conservation and water quality. It would also reduce fossil fuel use and emissions due to reduced pumping (Schaible and Aillery, 2012). As regards water quality, sediment and nutrient runoff decrease when irrigation is discontinued (Bjorneberg et al., 2002), but the magnitude depends on the runoff from the subsequent land use. Finally, subsidising water conservation in the form of improved irrigation efficiency and deficit irrigation may not always reduce water use and related emissions. Such actions also lower the costs of producing a crop and lead to an expansion of the cultivated land, thus increasing the amount of water used and decreasing the level of water quality related to the runoff of nutrients and pesticides (Pfeiffer and Lin, 2010; 2013).

Fertiliser management

Improving fertiliser management seeks to reduce nitrous oxide emissions from soil, in addition to reducing the carbon emissions associated with fertiliser manufacturing. This can involve reductions in fertiliser use, and thus excess applications, and improve plant uptake. The latter is achieved through improvements in application precision, use of time released forms, improving the timing of applications to better match timing of plant nutrient uptake, use of legumes in rotation or as winter cover crops, rotations with legumes, and use of nitrification inhibitors. Water quality gains occur through actions that reduce fertiliser use and associated nutrient runoff (Ongley, 1996). The relationship between fertiliser application and nutrient runoff is well-known, but is highly variable depending on local conditions and climatic events.

Chemical use reduction

Mitigation efforts may also reduce farm use of chemicals such as pesticides and lime. This would be done in an effort to reduce the GHG emissions generated in manufacturing those items and in an effort to increase productivity per hectare to limit total land required for production. Both would have water quality implications, and there might be minor water quantity implications due to the altered need for lime and replacement production. The water quality implications would largely be in the form of altered runoff of pesticide or lime residuals (Ongley, 1996; Hamilton and Helsel, 1995).

Manure management

Depending on the management system, manure can be a source of substantial methane emissions particularly when stored in a lagoon or some other wet environment. Such emissions can be altered through the use of digesters, covered manure lagoons or tanks, covered manure heaps, applying manure to crop lands, or combusting manure. In terms of water quality, application of manure to croplands increases the potential nutrient runoff and thus over application must be avoided (Ribaudo et al, 2003; Nielsen et al., 2012). If manure is incorporated into the soil, then immediate nutrient loss may be reduced by as much as 85% (Maguire et al., 2011). Use of best management application and storage practices would make a 40% difference in nutrient runoff (Young and Crowder, 1986). Digesters also reduce nutrient runoff into water (Vanotti et al., 2008).

Breeding and animal species choice

Greenhouse gas mitigation strategies may manipulate emissions through breeding or switching species. In terms of breeding, a number of scientists have suggested that it is possible to alter the enteric fermentation characteristics of animals by manipulating genetics, while others have indicated that increasing the retention time of food in the rumen would reduce the amount of nitrogen excreted. As regards species, switching from ruminants to non-ruminant livestock, such as hogs, broilers, turkeys and fish, would reduce enteric fermentation. Depending on the species, this could also lead to less feed per unit of meat produced (de Vries and de Boer, 2010, as shown for selected meats in **Figure C.1**) and produce different amounts of manure per unit of meat product produced.

Figure C.1. Land use for livestock products (in m²/kg of product)

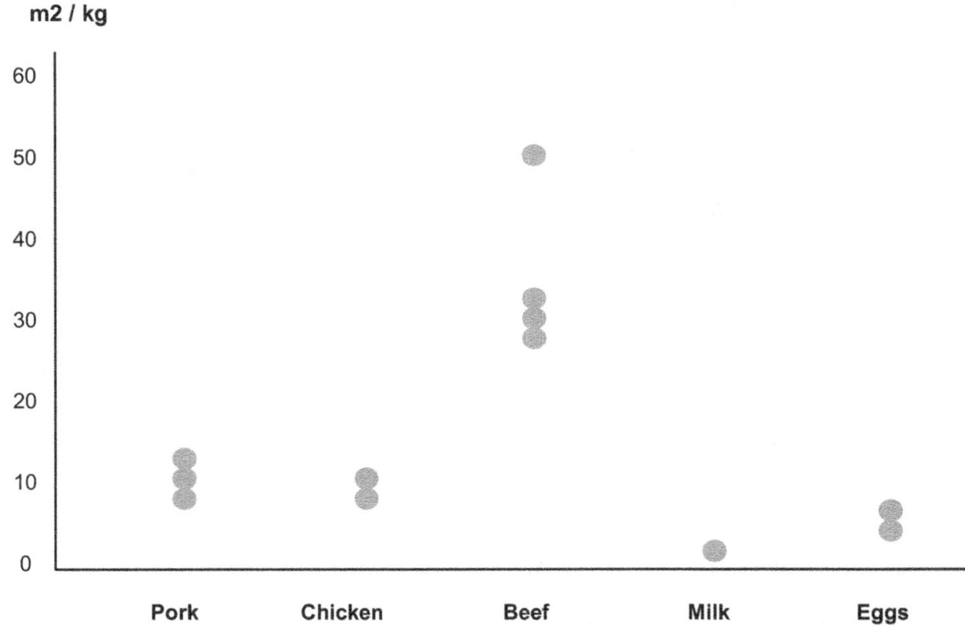

Source: de Vries, M. and I.J.M. de Boer (2010), "Comparing environmental impacts for livestock products: A review of life cycle assessments", *Livestock Science*, Vol. 128, Issues 1–3.

Changes in livestock species to ones that use less feed per unit of meat produced would reduce total feed demand, and associated water use and water quality effects. In terms of water quantity, there are large possible changes as indicated by the water footprint data in **Table C.1**. Note that such a water footprint is an average not a marginal measure and species switches may not change water use in such a pattern; for example, grassland use and associated water resource use may not shift with reductions in cattle herd size. Water quality would also be impacted by possible changes in the nitrogen and phosphorus content of excretions which could reduce nutrient runoff and improve water quality.

In addition, changes in the enteric fermentation characteristics could make it possible to alter the diet mix between roughage and feed concentrates that in turn affects land allocation and thus related water quantity and quality impacts.

Table C.1. Water footprint of selected meat and other products

Product	Per tonne product (m³/tonne)	Per calorie (litre/kcal)	Per unit protein (litre/gr protein)
Milk	1 020	1.82	31
Eggs	3 265	2.29	29
Chicken meat	4 325	3.00	34
Butter	5 553	0.72	0
Pork	5 988	2.15	57
Sheep/goat meat	8 763	4.25	63
Beef	15 415	10.19	112
Cereals	1 644	0.51	21
Pulses	4 055	1.19	19

Source: Adapted from Mekonnen, M.M. and A.Y. Hoekstra (2012), "A Global Assessment of the Water Footprint of Farm Animal Products", *Ecosystems*, Vol. 15.

Bioenergy

Bioenergy is a possible mitigation strategy that can involve substantial volumes of water (McCarl and Schneider, 2000, McCarl, 2008). The water quantity and quality effects of the increased pursuit of bioenergy depend on what kind of bioenergy processes and feedstock are used, as well as where the feedstock comes from. Bioenergy feedstocks can be used to generate liquid fuels like ethanol and biodiesel or to generate electricity. The processes involved in making these items are, for example, conventional ethanol fermentation, cellulosic ethanol production, feedstock combustion or the thermochemical conversion like pyrolysis.

Liquid fuels

Substantial amounts of water are used when making liquid fuels like ethanol (Gerbens-Leenesa et al., 2009; Higgins 2009). Aden (2007) reports that consumptive water use at a refinery is 3.5 to 4.0 times the volume of ethanol produced and that corn crop uses 785 units of water for every unit of ethanol produced (note that this figure should be considered against water consumption of the land used when the crop is not grown for ethanol). Water for refinery usage can be taken from groundwater with negative local effects. Water quality impacts arise due to fertiliser and pesticide inputs used in feedstock production, as discussed in the section on crop management (see reviews by CWIBP, 2008; Higgins, 2009).

Electricity

When producing electricity, substantial amounts of water can be used, mainly for feedstock production (Gerbens-Leenesa et al., 2009). The water footprint of bioelectricity is smaller than that of biofuels due to the more complete combustion of biomass. Water is also used at the generating facility, but this is largely for cooling purposes which are mostly non-consumptive. The water quality effects would largely come from the cultivation of feedstock; there could be a limited amount of combustion residuals having negative water quality effects under certain circumstances.

Pyrolysis – Biochar

Another bioenergy strategy involves pyrolysis where biomass feedstock is heated to the point they liquefy and generate bio oil, syngas and biochar (Lehmann and Joseph, 2009). Water quantity effects are largely due to feedstock cultivation and biochar effects. Biochar can be placed on land where it enhances nutrient and water holding characteristics and may reduce off-site runoff of water plus increase groundwater infiltration (Lehmann and Joseph, 2009). There are several possible implications for water quality. During the pyrolysis process, there is generation of water which would need to be carefully treated to avoid any detrimental water quality effects. Biochar can either be used in water treatment activities as activated charcoal or can be placed on land. In land application, it improves the nutrient holding characteristics and reduces nutrient runoff.

Feedstocks: Conventional crops and their residues

The use of conventional crops like corn or sugar cane as bioenergy feedstock would likely lead to an expansion in the acreage of conventional crops with an accompanying increase in water quantity (de Fraiture, et al 2008; Renouf et al., 2008) and quality effects as discussed under crop management above. It could also cause crop intensification and indirect land use change. The use of crop residues would remove surface cover from the land, potentially increasing erosion and nutrient runoff. On the other hand, removing residues of intensive crops such as corn can occur without causing much response and would allow expanded use of less intensive tillage (Wilhelm et al., 2004). This would likely improve water quality by reducing runoff, but would likely marginally decrease water quantity.

Feedstocks: Energy crops

The use of energy crops such as switch grass, hybrid poplar or miscanthus, would also have water quantity and quality impacts; this would vary based on the type of land. Use of conventional croplands would reduce erosion and nutrient runoff there, but would necessitate more land to replace lost production. Use of marginal lands would increase erosion and nutrient runoff by replacing generally lower input and more soil conserving alternatives. Jha et al. (2009) explore the water quality implications of producing alternative energy crops and find that a change from traditional crops to switchgrass would result in significant improvement of water quality.

Animal wastes

One can use manure as an input to biogas generation. Combustion would reduce the water quality effects of holding the manure in lagoons or otherwise disposing of it.

Processing by-products

The final crop category involves the use of processing by-products such as animal tallow, sugarcane bagasse, wheat milling residues, and vegetable oils. In these cases, water quality is improved through reduction of nutrient runoff.

Annex references

Aden, A. (2007), "Water Usage for Current and Future Ethanol Production", *Southwest Hydrology* September, pp. 22-23.

Beasley, R. P. (1972), "Erosion and sediment pollution control", *Iowa State University Press*, Ames, Iowa, USA, p. 320.

Bhardwaj, A.K., T. Zenone, P. Jasrotia, G.P. Robertson, J. Chen and S.K. Hamilton (2011), "Water and energy footprints of bioenergy production on marginal lands", *Global Change Biology Bioenergy*. 3(3), pp. 208-222.

Bjorneberg, D.L., D.T. Westermann and J.K. Aase (2002), "Nutrient losses in surface irrigation runoff", *Journal of Soil and Water Conservation* 57(524), pp. 524-529.

Brown, T.C., Binkley, D. (1993), "Forest practices as nonpoint sources of pollution in North America", *Water Resources Bulletin,* 29(5): pp. 729–740.

Clark, G.M., D.K. Mueller, and M.A. Mast. (2000), "Nutrient concentrations and yields in undeveloped basin of the United States", *Journal of the American Water Resources Association*, Vol. 36(4), pp. 849-860.

Committee on Water Implications of Biofuels Production in the United States, J.L. Schnoor, O.C. Doering III, D. Entekhabi, E.A. Hiler, T.L. Hullar, G.D. Tilman, W.S. Logan, N. Huddleston and M.L. Stoever (2008), *Water Implications of Biofuels Production in the United States*, National Research Council of the National Academies, The National Academies Press, Washington, D.C.

de Fraiture, C., M. Giordano and Y. Liao (2008), "Biofuels and implications for agricultural water use: blue impacts of green energy", *Water Policy*, Vol. 10, Supplement 1, pp. 67–81.

de Vries, M. and I.J.M. de Boer (2010), "Comparing environmental impacts for livestock products: A review of life cycle assessments", *Livestock Science*, Vol. 128, Issues 1–3, pp. 1-11.

FAO (2011), *Climate change, Water and Food Security*, FAO Water Report No. 36, Food and Agriculture Organization of the United Nations (FAO), Rome.

Fulton, S. and B. West (2002), "Forestry impacts on water quality". In: Wear, D., and J. Greis, (Eds.). Southern forest resource assessment. General Technical Report SRS-53, Asheville, NC: US Department of Agriculture, Forest Service, Southern Research Station, pp. 501–518.

Gerbens-Leenesa, W., A.Y. Hoekstraa and T.H. van der Meerb (2009), "The water footprint of bioenergy", *Proceedings of the National Academy of Sciences of the United States of America*, Vol. 106, No. 25, pp. 10219–10223.

Hamilton, P.A. and D.R. Helsel (1995), "Effects of Agriculture on Ground-Water Quality in Five Regions of the United States", *Ground Water*, Vol. 33, Issue 2, pp. 217–226.

Hansen, J., Mki. Sato, and R. Ruedy (2012), "Perception of climate change", *Proceedings of the National Academy of Sciences of the United States of America*, Vol. 109, 14726-14727, E2415-E2423.

Higgins, L.M. (2009), *Regional Differences in Corn Ethanol Production: Profitability and Potential Water Demands*, Unpublished PhD Dissertation, Department of Agricultural Economics, Texas A&M University.

Holland, J.M. (2004), "The environmental consequences of adopting conservation tillage in Europe: reviewing the evidence", *Agriculture, Ecosystems and Environment* Vol. 103, pp. 1–25.

Jha M., B.A. Babcock, P.W. Gassman, and C.L. Kling (2009), "Economic and environmental impacts of alternative energy crops", *International Agricultural Engineering Journal* 18(3-4), pp. 15-23.

Lehmann J. and S. Joseph (2009), *Biochar for Environmental Management: Science and Technology*, Earthscan, London.

Leterme, B. and D. Mallants (2011), "Climate And Land Use Change Impacts On Groundwater Recharge", *Proceedings ModelCARE2011*, Leipzig, Germany, September.

Maguire, R.O., P.J.A. Kleinman, C.J. Dell, D.B. Beegle, R.C. Brandt, J.M. McGrath, and Q.M. Ketterings (2011), "Manure Application Technology in Reduced Tillage and Forage Systems: A Review", *Journal of Environmental Quality*, Vol. 40, pp. 292–301.

Mayer, P.M., S.K. Reynolds, M.D. McCutchen and T.J. Canfield (2007), "Meta-Analysis of Nitrogen Removal in Riparian Buffers", *Journal of Environmental Quality*, Vol. 36, No. 4, pp. 1172-1180.

McCarl, B.A. (2008), "Bioenergy in a greenhouse gas mitigating world", *Choices*, 23(1), pp. 31-33

McCarl, B.A., and U.A. Schneider (2000), "US Agriculture's Role in a Greenhouse Gas Emission Mitigation World: An Economic Perspective", *Review of Agricultural Economics*, 22(1), pp. 134-159.

Mekonnen, M.M. and A.Y. Hoekstra (2012), "A Global Assessment of the Water Footprint of Farm Animal Products" *Ecosystems*, 15: 401–415.

Moldenhauer, W.C., G.W. Langdale, W. Frye, D.K. McCool, R.L. Papendick, D.E. Smika, and D.W. Fryrear (1983), "Tillage for Erosion Control", *Journal of Soil and Water Conservation*, Vol. 38, No. 3, pp. 144-151

Nielsen, A., D. Trolle, M. Søndergaard, T.L. Lauridsen, R. Bjerring, J.E. Olesen, and E. Jeppesen (2012), "Watershed land use effects on lake water quality in Denmark", *Ecological Applications*, 22(4), pp. 1187–1200.

Ongley, E. D. (1996), "Control of water pollution from agriculture", *FAO irrigation and drainage paper* 55, Food and Agriculture Organization of the United Nations, Rome.

Pfeiffer, L., and C.-Y.C. Lin (2013), "Does Efficient Irrigation Technology Lead to Reduced Groundwater Extraction? Empirical Evidence," unpublished paper, University of California, Davis.

Pfeiffer, L. and C.-Y.C. Lin (2010), "The effect of irrigation technology on groundwater use". *Choices*, Vol. 25 (3).

Pikul, J.L. and J.K. Aase (2003), "Water Infiltration and Storage affected by Subsoiling and Subsequent Tillage", *Soil Science Society of America Journal*, No. 67, pp. 859–866

Rabotyagov, S.S., M.K. Jha and T. Campbell (2010), "Impact of crop rotations on optimal selection of conservation practices for water quality protection", *Journal of Soil and Water Conservation*, Vol. 65, No. 6, pp. 369-380.

Renouf, M. A., M.K. Wegener and L.K. Nielsen (2008), "An environmental life cycle assessment comparing Australian sugarcane with US corn and UK sugar beet as producers of sugars for fermentation", *Biomass and Bioenergy*, Vol. 32, Issue 12, pp. 1144–1155.

Ribaudo M., N. Gollehon, M. Aillery, J. Kaplan, R. Johansson, J. Agapoff, L. Christensen, V. Breneman, and M. Peters (2003), "Manure Management for Water Quality: Costs to Animal

Feeding Operations of Applying Manure Nutrients to Land", USDA, Economic Research Service, *Agricultural Economic Report*, No. (AER-824), pp. 97.

Schaible G.D., Aillery M.P. (2012), "Water conservation in irrigated agriculture: trends and challenges in the face of emerging demands", *USDA ERS Economic Information Bulletin* No. 99.

Schnoor J.L., O.C. Doering III, D. Entekhabi, E.A. Hiler, T.L. Huller, G.D. Tilman, W.S. Logan, N. Huddleston and M. Stoever (2007), *Water Implications of Biofuels Production in the United States*. Washington, DC, USA: The National Academy of Sciences.

Wilhelm, W.W., J.M.F. Johnson, J.L. Hatfield, W.B. Voorhees and D.R. Linden (2004), "Crop and Soil Productivity Response to Corn Residue Removal", *Agronomy*, Vol. 96, No. 1, pp. 1-17.

Van Dijk, P.M., F.J.P.M. Kwaad and M. Klapwijk (1996), "Retention of Water and Sediment by Grass Strips", *Hydrological Processes*, Vol. 10, Issue 8, pp. 1069–1080.

Vanotti, M.B., A.A. Szogi and C.A. Vives (2008), "Greenhouse gas emission reduction and environmental quality improvement from implementation of aerobic waste treatment systems in swine farms", *Waste Management*, Vol. 28, Issue 4, pp. 759–766.

Weller, D.E., T.E. Jordan, D.L. Correll and Z.J. Liu (2003), "Effects of land-use change on nutrient discharges from the Patuxent River watershed", *Estuaries*, Vol. 26, Issue 2, pp. 244-266.

Wright, D.L., J.J. Marois, and T.W. Katsvairo (2012), "Transitioning from Conventional to Organic Farming Using Conservation Tillage in Florida", SS-AGR-11, Agronomy Department, Florida Cooperative EXtension Service, Institute of Food and Agricultural Sciences, University of Florida.

Young, C.E. and B.M. Crowder (1986), "Managing Nutrient Losses: Some Empirical Results on the Potential Water Quality Effects", *Northeastern Journal of Agricultural and Resource Economics*, Vol. 15, No. 2, pp 130-136.

ORGANISATION FOR ECONOMIC CO-OPERATION AND DEVELOPMENT

The OECD is a unique forum where governments work together to address the economic, social and environmental challenges of globalisation. The OECD is also at the forefront of efforts to understand and to help governments respond to new developments and concerns, such as corporate governance, the information economy and the challenges of an ageing population. The Organisation provides a setting where governments can compare policy experiences, seek answers to common problems, identify good practice and work to co-ordinate domestic and international policies.

The OECD member countries are: Australia, Austria, Belgium, Canada, Chile, the Czech Republic, Denmark, Estonia, Finland, France, Germany, Greece, Hungary, Iceland, Ireland, Israel, Italy, Japan, Korea, Luxembourg, Mexico, the Netherlands, New Zealand, Norway, Poland, Portugal, the Slovak Republic, Slovenia, Spain, Sweden, Switzerland, Turkey, the United Kingdom and the United States. The European Union takes part in the work of the OECD.

OECD Publishing disseminates widely the results of the Organisation's statistics gathering and research on economic, social and environmental issues, as well as the conventions, guidelines and standards agreed by its members.

OECD PUBLISHING, 2, rue André-Pascal, 75775 PARIS CEDEX 16
(51 2014 03 1 P) ISBN 978-92-64-20912-1 – 2014-1

IWA Publishing's authorised EU representative for General Product
Safety Regulations is Diane D'Arras, 15 rue Duret, 75116 Paris,
France, e-mail: safety@iwap.co.uk.

Printed and bound by CPI Group (UK) Ltd, Croydon, CR0 4YY
12/05/2026
02108892-0007